NIJINSKY
THE FILM

NIJINSKY

THE FILM

Text by

ROLAND GELATT

BALLANTINE BOOKS · NEW YORK

GRAPHIC CREDITS

This book was photo-composed by Haber Typographers, Inc., New York, New York.
The full-color separations and special Duotone reproductions were supplied by Creative Lithographers, New York, New York.
The book was printed by Connecticut Printers, Inc., Bloomfield, Connecticut.
The binding was done by A. Horowitz & Sons, Clifton, New Jersey.
The special foil cover was printed by Album Graphics, Inc., Melrose Park, Illinois.

Deborah Speed designed the book.
James R. Harris designed the cover.
Barry Larit directed the production and manufacturing.

Special photographers: Dennis Cameron, Michael Childers, Zoë Dominic.

Photograph of Vaslav and Romola Nijinsky, Vienna, 1945, by John Phillips, Life Magazine © 1945 Time, Inc.

The publishers gratefully acknowledge the cooperation of Alfred A. Knopf, Inc. in reproducing historical photographs from Nijinsky Dancing by Lincoln Kirstein.

Published in the United States by Ballantine Books, a division of Random House, Inc., New York
and simultaneously in Canada by Random House of Canada, Limited, Toronto, Canada.

Library of Congress Catalog Card Number: 80-10466
ISBN 0-345-28899-8
ISBN 0-345-28607-3 pbk.

Manufactured in the United States of America
First Edition: April 1980

PARAMOUNT PICTURES PRESENTS

A HARRY SALTZMAN PRODUCTION • A HERBERT ROSS FILM

STARRING

Alan Bates • Leslie Browne • and George de la Peña, "Nijinsky"

ALSO STARRING

Alan Badel • Colin Blakely • Carla Fracci • and The London Festival Ballet
Music Performed by The Los Angeles Philharmonic • Adapted and Conducted by John Lanchbery
Additional Choreography by Kenneth MacMillan • Associate Producer Howard Jeffrey
Executive Producer Harry Saltzman • Screenplay by Hugh Wheeler
Produced by Nora Kaye and Stanley O'Toole • Directed by Herbert Ross
A Paramount Picture

Vaslav Nijinsky as the Golden Slave in <u>Schéhérazade</u>

WE will say of a new dancer who comes along that he is talented but no Nijinsky. Such is the stature and resonance of Diaghilev's first and foremost star that we still intone his name—more than 60 years after his last public appearance—as the standard by which all aspirants are measured. Along with Caruso and Garbo and Chaplin, Nijinsky belongs in that select company of legendary idols whose preeminence is taken as axiomatic.

His reputation rests, nevertheless, on the flimsiest of foundations. Of Nijinsky's abilities as a dancer we have only the testimony of eye-witnesses and the frozen images of a few photographs. Of his achievements as a choreographer the record is even more nebulous. And yet, even from the imperfect evidence that survives, there can be no doubt whatever that Nijinsky raised the art of dance to wholly new levels and turned the development of ballet in fresh and fruitful directions.

If that were all, he would still be a figure to conjure with. But there is more to the Nijinsky legend than balletic mastery alone. For us in the age of anxiety, his strange and unsettling life story holds a particular pertinence. He was a loner, sexually ambivalent, the prototype of alienated man, who dazzled the world for a few glorious years only to retreat into the curtained world of schizophrenia midway in his life. If anyone in this century can be said to have a valid claim to the title of tragic genius, it is Vaslav Nijinsky.

His art, alas, was never captured in motion pictures. As if in compensation for that absurd oversight, the dream of re-creating the dancer's life on film has haunted producers, directors, and actors for more than 40 years. That dream for long remained just a dream. Every attempt to produce a film about Nijinsky became mired in a tangle of problems and frustrations.

Now at long last the story of Nijinsky has reached the screen. This book is about the making of Herbert Ross's film. But before we turn to its gestation, a look at the "real" Nijinsky is in order. Certain aspects of the dancer's career are only hinted at in the film, which confines itself to a period of twenty months, from March 1912 to December 1913. Within that brief interval were concentrated the salient and climactic episodes of Nijinsky's life. To see these events in full perspective, however, we must go back to the beginning.

Nijinsky was born in the right place at the right time. Russia at the close of the nineteenth century was an ideal spawning ground for a dancer and choreographer of genius. By then, ballet had had a history of some 300 years. From its courtly beginnings in Renaissance Italy the art of dance soon emigrated to France, where it moved from the ballroom into the theater and acquired a set of basic principles from the influential teacher and ballet master Jean-Georges Noverre. The vogue for ballet gradually spread to all the great capitals of Europe, but Paris retained artistic primacy for some two centuries. Its apogee was probably reached in the 1830s and 1840s, when ballerinas like Marie Taglioni, Fanny Elssler, and Carlotta Grisi vied for preeminence in such still-durable creations as La Sylphide and Giselle. Thereafter, a period of decadence afflicted the art of dance in Paris, and the center of balletic creativity shifted to St. Petersburg, capital of Imperial Russia.

The Imperial School of Ballet had been established in St. Petersburg in the eighteenth century with a corps of teachers recruited from Western Europe. This school, and the ballet company at the Maryinsky Theater which it nourished, became the great pride of Russia's aristocracy and intelligentsia. Vast sums were spent to make Russian ballet the greatest in the world, and by the 1870s its supremacy was unchallengeable. The accomplishment could be attributed in large part to three resident foreigners: Marius Petipa, a Frenchman, who was the Maryinsky's chief choreographer (and to whom we owe Swan Lake and Sleeping Beauty);

Christian Johannsen, a Dane, who reigned for 30 years as principal teacher, and who stood in the direct line of pedagogical descent from the classic French school of Noverre and Vestris; and Enrico Cecchetti, an Italian dancer and (later) ballet Master, whom we shall meet in the film. It was to this prestigious and richly endowed institution that the ten-year-old Vaslav Nijinsky applied for admission on August 20, 1898.

He was the second of three children born to Thomas and Eleonora Nijinsky, both Polish and both dancers. The family led the life of nomads as Thomas journeyed from one engagement to the next across the vast extent of Imperial Russia. He happened to be performing in Kiev when Vaslav was born there on March 12, 1888. Some books, notably Romola Nijinsky's biography, give 1890 as the date of the dancer's birth; but this was rather the year of his baptism, postponed for two years in order that it could be performed in Warsaw at a Roman Catholic church. Vaslav would learn some Polish from his mother, but the tongue in which he was most fluent, and the culture with which he most closely identified, was that of Russia. His older brother, Stanislav, was mentally retarded, perhaps as the result of a severe fall at the age of six, perhaps because of a hereditary failing. As we shall see in the movie, Vaslav feared the latter, and Stanislav's fate held a very real terror for him. The youngest child, Bronislava, was born in Minsk in 1891. She would in time become her brother's collaborator and later a distinguished choreographer in her own right.

Not long after Bronislava's birth, Thomas and Eleonora separated. Left on her own with three young children to support, the mother settled in St. Petersburg and took in boarders. It was a penurious existence from which there was only one hope of escape: for Vaslav to enter the Imperial

Nijinsky in uniform of the Imperial School of Ballet at about age fourteen (left) and eighteen (right)

School of Ballet and train for a career at the Maryinsky Theater. As we have seen, he applied for admission and was accepted as a pupil in 1898. Almost at once he showed the promise of a potentially great dancer, and in his early teens began to appear in small roles at the Maryinsky. One of these was that of the mulatto slave in Petipa's ballet <u>Le Roi Candaule</u>, in which the eighteen-year-old student danced during the Christmas season of 1906. His boyish grace illuminates a series of photographs made at the time. They contrast sharply with a picture of Nijinsky in street clothes, taken for his graduation the following spring. Here we can make out the unprepossessing teen-ager described by Alexandre Benois—"a short, rather thick-set little fellow with the most ordinary, colorless face! He was more like a shop assistant than a fairy-tale hero."

On stage, however, Nijinsky was wholly transformed. After his graduation performance in April 1907, the Maryinsky's prima ballerina, Matilda Kchessinskaya, came backstage to offer congratulations and to request him as her partner. Kchessinskaya's talents as a dancer were surpassed only by her conquests among the Romanovs; she had been the mistress of the Czar and was now the lover of the Grand Duke Sergei Mikhailovitch. Her patronage meant that all doors would be open to the nineteen-year-old prodigy.

Nijinsky was now freed from the strict and spartan discipline of school. From May to September, when the Maryinsky was closed, the young prodigy could do as he pleased. Even during the remaining eight months of the year, there were many evenings when he was not obliged to dance. This newfound freedom went to Nijinsky's head. He was often to be seen in the entourage of this or that aristocrat or millionaire, burning the candle at both ends.

From Richard Buckle's invaluable biography, we learn that Vaslav took an interest at this time in two girls from the Maryinsky corps de ballet—flirtations that take on particular significance in the light of subsequent events. But it was with a well-known homosexual, Prince Paul Dmitrievitch Lvov, that Nijinsky formed his first close relationship. For several months they were seen together at all the best places. Lvov soon disappeared from the young dancer's life, but not before effecting an introduction that would have long repercussions on the history of ballet. For it was through him that

Nijinsky met the remarkable man who would provide a focus to his life and set him on the course to immortality.

Sergei Pavlovitch Diaghilev was Nijinsky's elder by sixteen years. Born in 1872, of an aristocratic family, he had already made his mark in St. Petersburg as a catalyst, organizer, and patron of the arts. He had edited and published an influential magazine devoted to acquainting his fellow Russians with the latest artistic currents at home and abroad, and he had organized a spectacular exhibition of historic Russian portrait paintings gathered from

Diaghilev painted by Bakst, c. 1905

palaces and country estates throughout the Czar's domain. But it was as a self-appointed minister plenipotentiary for bringing the arts of Russia to the outside world that his extraordinary abilities came into full play. As a young man Diaghilev had traveled throughout Western Europe and—like so many of his countrymen—looked upon Paris as his second home. In 1906 he established a Russian artistic beachhead there by bringing his exhibition of Russian portrait paintings to the Salon d'Automne in the Petit Palais. A year later he organized a series of concerts at the Paris Opéra consecrated to the work of Russian composers from Glinka to Scriabin. These were followed in 1908 with a production in Paris of Mussorgsky's opera <u>Boris Godunov</u>, introducing the incomparable Feodor Chaliapin in his most celebrated role.

The acclaim that greeted this last Parisian venture was still the talk of St. Petersburg when Diaghilev and Nijinsky first met in November or December 1908. Diaghilev had come rather late in life to an appreciation of ballet. His first loves had been music (especially opera), painting, and poetry. But Alexandre Benois, the artist and stage designer, had convinced him that ballet was an art form worthy of serious attention, and Diaghilev had been further persuaded by the reforms of a young choreographer at the Maryinsky, Mikhail Fokine, who was intent on transforming ballet from mere graceful entertainment into a more involving dramatic expression. Fokine's talent had been somewhat hobbled by the Maryinsky's hidebound traditions, and Diaghilev sensed the breakthrough that he might achieve if liberated from its restrictions. Diaghilev also sensed the extraordinary impact that dancers like Pavlova, Karsavina, Kchessinskaya, and Nijinsky would make on the unsuspecting Parisian public.

Throughout that winter Diaghilev worked on

the forthcoming ballet season with his collaborators: Benois, Fokine, the painter Leon Bakst, the composer Igor Stravinsky, and other friends and advisers. His genius for marshaling great talent was coming into full bloom. While the ballets for Paris were being elaborated and designed, Diaghilev began to engage his dancers. At this point the undertaking almost foundered. When Matilda Kchessinskaya learned that she was scheduled to dance only one role in Paris, she withdrew and persuaded the Imperial Secretariat to cancel the large subsidy that had been promised. Diaghilev carried on despite this setback, raising money on his own in Russia and Paris. Circumstances were forcing him into the role of an entrepreneur. Nevertheless, he objected to being called an impresario. "I am not a professional impresario," he once said; "my specialty is to make painters, musicians, poets, and dancers work together."

At this juncture Nijinsky was still just one of several talented Maryinsky soloists engaged to appear in Paris, not yet the troupe's undisputed star.

Diaghilev with Stravinsky, 1921

ment. The critic Henri Gauthier-Villars, Colette's husband, rushed into print after the opening performance to proclaim his admiration for this "wonder of wonders." "Yesterday when Nijinsky took off so slowly and elegantly, describing a trajectory of 4-1/2 meters and landing noiselessly in the wings, an incredulous 'Ah!' burst from the ladies." This kind of adulation marked a new departure. "Since the Romantic period," Richard Buckle reminds us, "it had been the woman, the Muse, the diva, the ballerina who had been worshipped: to admire a man for his grace and beauty was unheard-of."

The admiration grew more worshipful as the season progressed. By the time it drew to a close, on June 19, le tout Paris was in thrall to the Russian ballet. By that time, too, the liaison between Diaghilev and Nijinsky had become an accomplished fact. Their attachment, which had been growing

With the close of the St. Petersburg season on May 1, Nijinsky and his colleagues set off by train for the French capital. Less than three weeks later, on May 19, 1909, Paris got its first glimpse of the Russian dancers. It was a case of instant adoration. The greatest revelation turned out to be the vitality and virtuosity of the company's male soloists. Dancing like this had never been seen in the West. Nijinsky in particular elicited gasps of astonish-

since the previous winter, was sealed at the very end of the Paris season, when Nijinsky fell desperately ill with typhoid fever. Diaghilev rented a furnished flat and took charge of nursing the patient back to health. From that moment they were inseparable.

To play the role of psychologist at a distance of seven decades is an exercise fraught with danger. We know that Diaghilev already had experienced a long history of homosexual relationships, that his love affairs with young men were to continue to the end of his life. Charming and gallant though he could be to women, there is no hint that he was ever sexually attracted to them. Nijinsky, on the other hand, would appear to have been bisexual. Certainly, events were to show that he was by no means an ingrained homosexual. But at the moment when his liaison with Diaghilev began, the older man seems to have filled a very real emotional need—that of a surrogate father who was to provide not only a mantle of security but also an unerring sense of artistic direction. Initiated out of mutual desire, the relationship blossomed into a creative partnership of far-reaching consequences.

Back in St. Petersburg, while Nijinsky resumed his duties at the Maryinsky, plans for Diaghilev's second Paris season were set in motion. Once again acute financial problems imperiled the enterprise. This time Diaghilev was rescued by an eccentric and engaging philanthropist who plays an important part in the film: Baron Dmitri de Gunzburg. The scion of a wealthy Jewish banking family, harboring a fervent passion for ballet, he took it upon himself to serve as Diaghilev's chancellor of the exchequer with a business sense that could most charitably be described as bizarre. The baron kept accounts on the cuffs of his shirts and invariably seemed surprised when these essential figures disappeared in the laundry.

For the 1910 Paris season, Diaghilev had commissioned three new Fokine ballets: <u>Firebird</u> (with a score by the still unknown Stravinsky), <u>Carnaval</u>, and <u>Schéhérazade</u>. They helped to make this second visit an even more thundering success than the first. Nijinsky's spectacular virtuosity again held audiences spellbound. In <u>Schéhérazade</u>—with its opulent sets and costumes designed by Bakst, its evocation of a distant and exotic culture, its titillating final orgy—he seemed the very epitome of primitive abandon. "Nijinsky shot out of his room," Cyril Beaumont reports, "like an arrow from a bow in a mighty parabola which enabled him to cross in one bound a good two-thirds of the width of the stage." This description of the Golden Slave's entrance makes a more compelling case for Nijinsky's impact in the role than the stagey and earthbound photographs made in a London studio by Baron Adolphe de Meyer a year after the Paris premiere.

Nijinsky as the Golden Slave, photographed by De Meyer

Exhilarated by the immense success of his second ballet season, Diaghilev now made a bold decision. Up to this time his company had consisted of an <u>ad hoc</u> troupe of dancers on summer leave from the permanent companies in St. Petersburg and Moscow. This meant that rehearsals and performances had all to be crammed within the ten-week period between May 1, when the Russian companies disbanded for the summer, and July 14, when Parisian high society fled to the mountains or the seashore. "It seems senseless," Diaghilev wrote to his associates in the summer of 1910, "to go on assembling a fresh company every year only to perform in Paris for a short season. Our very success proves that abroad there is a demand for ballet and that we should be all but certain to succeed. After taking everything into consideration I propose founding for the first time a large private company." With such a company Diaghilev hoped to tour the capitals of Europe for most of the year.

There was one major obstacle. Graduates of the Imperial Ballet School were obligated to dance at the Maryinsky for a minimum of five years after leaving school. In Nijinsky's case this meant that he would be unavailable on a year-round basis until 1912 at the earliest. As he was now Diaghilev's most acclaimed star, to say nothing of being his constant companion, it was unthinkable that Nijinsky should be immured in the Maryinsky.

Did events solve this problem for Diaghilev of their own accord, or did he manipulate events to provide a solution? We shall probably never know. In any case, we do know that when Nijinsky danced in <u>Giselle</u> for the first time at the Maryinsky, in January 1911, his costume caused grave offense. He had danced the role of Albrecht in Paris the previous spring and had worn a costume specially designed by Benois for that production. Now in

Nijinsky in <u>Giselle</u>. Right: with Karsavina

St. Petersburg he insisted on wearing it again. To our eyes, Benois's costume hardly seems very daring. But to Russian eyes of seventy years ago, it appeared excessively revealing. The dancer refused to have it altered in any way, and next day he was given an ultimatum: apologize or resign. He resigned. We can draw our own conclusions whether Diaghilev was coaching from the sidelines.

Les Ballets Russes de Diaghilev, as the new permanent company would hence be known, made its debut on April 9, 1911, at the opera house in Monte Carlo. Diaghilev had brought his dancers there from all over Europe, and as their coach he had imported Enrico Cecchetti, the Maryinsky's veteran teacher, who would now be regularly attached to the Diaghilev ballet. Every day, as we shall see in the film, Cecchetti gave class to the entire company. "Those wonderful lessons, which began from the day of arrival [in Monte Carlo], were not only of enormous assistance to everyone, but at once imposed a new style and attitude on the dancers not drawn from the Imperial Theaters and were a boon to Fokine by welding the company into a whole." The comment comes from Serge Grigoriev, another person we shall encounter in the film, who served as Diaghilev's faithful stage manager throughout the history of the Ballets Russes.

Fokine's Le Spectre de la Rose, a delicate invention set to Weber's Invitation to the Dance, had its first performance during this debut season in Monte Carlo. It was to become the most popular piece in the Diaghilev repertory and provided Nijinsky with his most celebrated role. As the fleeting incarnation of a young girl's reveries, Nijinsky gave "a perfect illusion of something non-human, something apart from the earth and sex and life, the spirit not of a human being but of a flower," according to Prince Peter Lieven. The moment for bated breath came when the dancer made his final leap through an open window, "a jump so

Nijinsky in <u>Le Spectre de la Rose</u>. Right: with Karsavina

De la Pena as Nijinsky in <u>Le Spectre de la Rose</u>

poignant, so contrary to all the laws of flight and balance, following so high and curved a trajectory, that I shall never again smell a rose without this ineffacable phantom appearing before me," according to Jean Cocteau. Perhaps nothing so astonished Nijinsky's contemporaries as his capacity for soaring into the air and giving the illusion of hovering motionless at the top of his rise. Tamara

Karsavina, Diaghilev's prima ballerina, who partnered Nijinsky in Le Spectre and was always to be a close and loyal friend, refers in her autobiography to this legendary leap:

> Somebody was asking Nijinsky if it was difficult to stay in the air as he did while jumping; he did not understand at first, and then very obligingly: "No! No! Not difficult. You have to just go up and then pause a little up there."

Left: De la Pena and Fracci in Le Spectre

When the company reached Paris in June it was ready to give birth to another of Fokine's finest creations, the ballet <u>Petrouchka</u>. Set to Stravinsky's exuberant and scintillating music, with decor by Benois, <u>Petrouchka</u> exalted the popular culture of Russia in a series of vivid tableaux. In his role as the rueful and ungainly puppet, Nijinsky achieved a miracle of transmutation. This ability to enter the very soul of a character was perhaps the dancer's greatest gift. "When he put on his costume," Benois recalls, "he gradually began to change into another being, the one he saw in the mirror. He became reincarnated and actually entered into his new existence." To portray an anti-hero like Petrouchka was something no leading dancer had ever been asked to do before. "I was surprised," Benois goes on to say, "at the courage Vaslav showed, after all his <u>jeune premier</u> successes, in appearing as a horrible half-doll, half-human grotesque. The great difficulty of Petrouchka's part is to express his pitiful oppression and his hopeless efforts to achieve personal dignity <u>without ceasing to be a puppet</u>. Both music and libretto are spasmodically interrupted by out-

Nijinsky as Petrouchka.   Right: De la Pena as Petrouchka

bursts of illusive joy and frenzied despair. The artist is not given a single <u>pas</u> or <u>fioriture</u> to enable him to be attractive to the public, and one must remember that Nijinsky was then quite a young man and the temptation to be 'attractive to the public' must have appealed to him far more strongly than to an older artist." But of course the point is that by 1911 Nijinsky had already made the transition from feted prodigy to that of serious artist.

Diaghilev's company now began to make the rounds of Europe. Perhaps nowhere did they create so indelible an impression as in London, which first succumbed to the splendors of the Ballets Russes during the festive celebrations surrounding the coronation of King George V in the summer of 1911. Encomiums were lavished on the Russian dancers, and as usual the highest praise was reserved for Nijinsky. "His every movement is instinct with spontaneity and grace," rhapsodized the <u>Sunday Times.</u> "He seems to be positively lighter than air, for his leaps have no sense of effort and you are inclined to doubt if he really touches the stage between them."

Despite the panegyrics of critics and the resultant clamor for tickets, Diaghilev had always to find wealthy patrons to make up his deficits. Running the world's most prestigious ballet company was never to be a paying proposition. In the film we are shown the kind of elaborate receptions that Diaghilev had to attend—usually with Nijinsky— in order to meet and flatter potential backers. One of the most striking of these scenes takes place at the country house of the Marchioness of Ripon. Here, in the twilight of a summer evening, Nijinsky is seen dancing Harlequin's solo from <u>Carnaval</u> for the private delight of her guests. This formidable <u>grande dame</u> and her daughter Lady Juliet Duff were to be of inestimable help in smoothing the way for Diaghilev's London seasons.

Nijinsky as Harlequin in <u>Carnaval</u>

We now come to a conflagration that indirectly changed the course of Nijinsky's life. Diaghilev had arranged for the Ballets Russes to give a five-week season at the Narodny Dom in St. Petersburg early in 1912. At last his compatriots were to have a chance to experience at first hand the new company and the new repertory that had so bedazzled audiences in the West. But it was not to be. In January the Narodny Dom burned to the ground, and no other suitable theater in St. Petersburg could be found. Diaghilev's disappointment at depriving St. Petersburg of the Ballets Russes (forever, as it turned out) was overshadowed by the need to find alternative engagements for his company. He went into action and secured last-minute bookings in three Central European cities. One of them was Budapest.

It is tempting to speculate on the possible turn of events had the company gone instead, say, to Prague. Would Nijinsky and Diaghilev eventually have fallen out just the same? Would the dancer still have married? Would he have gone insane? The questions are as intriguing as they are unanswerable. We are left with what actually happened. On March 5, 1912, the Ballets Russes opened a one-week engagement at the Budapest Opera House. In the audience was Romola de Pulsky.

No one would ever fall with such headlong abandon into the ecstasies of balletomania as this pampered 21-year-old. Before the week was up, Romola de Pulsky had resolved to become a dancer, join the Ballets Russes, and meet the great Nijinsky. In time she succeeded in accomplishing all that and more.

The Herbert Ross film begins during this fateful week in Budapest. At that time, of course, Diaghilev and Nijinsky were still totally unaware of Romola's existence. Their attention was entirely focused on a new ballet set to Debussy's L'Après-midi d'un Faune, which Vaslav was about to choreograph. Mikhail Fokine, the company's chief choreographer from the very first Paris

season, had begun to fall from favor. "Just as Fokine had rebelled against the academic dance, throwing out tutus, turn-out, and virtuosity for its own sake," Richard Buckle explains, "so were the two friends feeling for different reasons dissatisfied with the results of Fokine's revolution.... Diaghilev foresaw a dead-end to the ballet of local color and the evocation of past periods or distant lands, and he had a prejudice against stories and drama in ballet....Better than to evoke past eras would surely be to re-interpret them or even to speak for your own." He was not alone in his criticism of Fokine. Writing to his mother in 1912, Stravinsky refers to Fokine as "an exhausted artist, one who has traveled his road quickly, and who writes himself out with each new work....

New forms must be created." It was to his young protégé Nijinsky that Diaghilev entrusted this formidable challenge.

L'Après-midi d'un Faune was indeed a break with the past. "Its choreography was not choreography as we understood the term," Grigoriev writes. "The dancers merely moved rhythmically to the music and then stopped in attitudes, which they held. Nijinsky's aim was, as it were, to set in motion an archaic Greek bas-relief, and to produce this effect he made the dancers move with bent knees and feet placed flat on the ground heel first (thereby reversing the classical rule). They had also to keep their heads in profile while still making their bodies face the audience, and to hold their arms rigid in various angular positions." The first

Far right: De la Pena as Nijinsky

Nijinsky in <u>Faune</u>

Far left: De la Pena as Nijinsky with Valerie Aitken as the First Nymph

performance in Paris, on May 29, 1912, provoked Diaghilev's first full-fledged scandal. At the end there were as many boos as shouts of approval. The fact that Nijinsky's ballet ends with an act of simulated masturbation did nothing to appease the outrage of those who had already found the dancing static and angular. Was this final erotic gesture a last-minute improvisation by Nijinsky? The film, drawing on a suggestion in Prince Peter Lieven's book, depicts it as having been wholly unpremeditated, but according to Grigoriev the scene was rehearsed that way from the start.

In the ensuing controversy that broke out in the press, no less an eminence than the aged sculptor Rodin sprang to Nijinsky's defense. His article provides a vivid eye-witness account of the occasion:

Nijinsky has never been so remarkable as in his latest role. No more jumps—nothing but half-conscious animal gestures and poses. He lies down, leans on his elbow, walks with bent knees, draws himself up, advancing and retreating, sometimes slowly, sometimes with jerky angular movements. His eyes flicker, he stretches his arms, he opens his hands out flat, the fingers together, and as he turns away his head he continues to express his desire with a deliberate awkwardness that seems natural. Form and meaning are indissolubly wedded in his body, which is totally expressive of the mind within....His beauty is that of antique frescoes and sculpture: he is the ideal model, whom one longs to draw and sculpt. When the curtain rises to reveal him reclining on the ground, one knee raised, the pipe at his lips, you would think him a statue; and nothing could be more striking than the impulse with which, at the climax, he lies face down on the secreted veil, kissing it and hugging it to him with passionate abandon....

At least two novelties would be needed for the Paris season of 1913, and Nijinsky was expected to provide them. Diaghilev was not in the least dismayed at the prospect of another <u>succès de scandale.</u> All the commotion surrounding <u>L'Après-midi d'un Faune</u> had kept the box office working overtime. But little could Diaghilev have guessed the uproar that would greet <u>Le Sacre du Printemps.</u>

Stravinsky's cataclysmic score, with its complex rhythms and constantly changing meter, posed an enormous challenge. To meet it, Nijinsky dreamed of originating a new kind of dance, as precedent-shattering as the music itself. He knew what he wanted, or thought he knew what he wanted, but found himself unable to articulate his ideas in a way that colleagues could understand. The spasmic and frenzied movements he envisaged ran counter to everything his dancers had been taught. Frustrated and nonplussed, he would rage at their seeming lack of comprehension and dedication. Diaghilev happened to walk into the rehearsal room during one of these tantrums and furiously rebuked Nijinsky in front of the company. This time of travail, well documented by those who suffered through it, is carefully reconstructed in the film.

Meanwhile, the always touchy Fokine—nominally the company's choreographic director—was complaining bitterly about Nijinsky's intrusion into his domain. "I had a violent argument with Diaghilev," he recounts in his memoirs. "I used words which described his relationship with Nijinsky in plain terms. I shouted that the ballet company was turning from a fine art into a perverted degeneracy." The film depicts his mounting resentment, culminating in a stormy confrontation with Diaghilev during the course of a rehearsal. This outburst marked the end, for a while at least, of Fokine's association with the Ballets Russes. With his departure, Nijinsky's choreographic responsibilities suddenly became more pressing.

While rehearsals for <u>Le Sacre</u> dragged on, Nijinsky cobbled together at the last minute another and slighter ballet, in which for the first time he appeared (along with Karsavina and Ludmilla Schollar) in contemporary dress. This was <u>Jeux</u>, set to a new score commissioned from Debussy. It opened the 1913 season of the Ballets Russes in Paris to damp response. The ballet soon disappeared from the repertory, and only a few postured photographs and some drawings remain.

Nijinsky in <u>Jeux</u>. Above center: De la Pena as Nijinsky

Right: De la Pena as Nijinsky with Carla Fracci and Genesia Rosato, above center, below left and right

Two weeks later, on May 29, 1913, came the riotous first performance of <u>Le Sacre du Printemps.</u> Valentine Gross, whose sketches of the ballet provide one of the keys to Kenneth MacMillan's reconstruction of Nijinsky's choreography, sat in the midst of the fray. "The theater seemed to be shaken by an earthquake. People shouted insults, howled and whistled, drowning the music. There was slapping and even punching....I cannot think how it was possible for this ballet, which the public of 1913 found so difficult, to be danced through to the end in such an uproar." Stravinsky's "cacophonous" music in combination with Nijinsky's "epileptic"

choreography had detonated one of the loudest explosions in theatrical history. Did Diaghilev begin to question his good judgment in replacing Fokine with someone as inexperienced as Nijinsky? The film suggests that he did. But on the night of the first performance, there seem to have been no misgivings. "I went with Diaghilev and Nijinsky to a restaurant," Stravinsky reports. "So far from weeping and reciting Pushkin in the Bois de Boulogne as the legend is, Diaghilev's only comment was: 'Exactly what I wanted'. He certainly looked contented. No one could have been quicker to understand the publicity value."

Dancers from the original Ballets Russes production of <u>Le Sacre du Printemps</u>

De la Pena as Nijinsky choreographs <u>Le Sacre</u>

Monica Mason as the Chosen Virgin in <u>Le Sacre du Printemps</u>

But was Vaslav wholly contented? In the light of after events, it seems unlikely. For four years he had played Trilby to Diaghilev's Svengali, the androgynous puppet of a master manipulator, constantly on the move through the world of posh hotels and posh society, totally immersed in the problems and politics of the Ballets Russes. He had professional colleagues but no real friends, for his special relationship with the head of the company effectively isolated him from comradeship with others. In 1909, fresh from the Imperial Ballet School, untraveled and unsophisticated, floundering in the fleshpots of St. Petersburg, the young Nijinsky had welcomed Diaghilev's firm guidance. At the Maryinsky Theater he had been merely a promising talent; the Ballets Russes made him the world's most acclaimed dancer. But now, exhausted from giving birth to three pathbreaking and unappreciated ballets, did he long for some escape? Did he perhaps even begin to look with more than passing interest at attractive young women?

The sequence of events in the immediate aftermath of the Sacre premiere is clear. What passed between Diaghilev and Nijinsky is not. As usual, the company went on to London from Paris (in all, Le Sacre was performed seven times that season, only once to general pandemonium). The troupe then had a short respite before it was to sail, in mid-August, for a tour of South America. Diaghilev and Nijinsky spent the interval in Baden-Baden, where Benois joined them to discuss plans for a new ballet set to the music of Bach. Benois later wrote to his friend Stravinsky: "I saw Sergei and Vaslav almost on the eve of Vaslav's departure for Argentina, and there was no hint then about the coming event. Nijinsky was attentively studying Bach with us, preparing the Bach ballet." But that lay on the surface. The film suggests what was brewing beneath. Whatever happened, and for whatever reason, we know that Diaghilev decided at the last minute to remain behind in Europe. Baron de Gunzburg would serve as his deputy on the South American tour.

Romola had meanwhile succeeded in attaching herself to the Ballets Russes. With the advantage of good connections (her mother, Emilia Markus, was Hungary's leading actress, her brother-in-law a celebrated tenor in Vienna) and of family money, she persuaded Diaghilev and Maestro Cecchetti to let her tag along as a fee-paying student dancer. Now, with the rest of the company, she embarked on the Royal Mail Line's S.S. Avon for Buenos Aires. Unlike others in the corps de ballet, she booked passage in First Class.

Emilia Markus

What transpired during the course of that three-week voyage was so unlikely, so unexpected, that even those closest to the scene could hardly give it credence. Diaghilev's remote and cocooned paramour had struck up a shipboard romance with Romola de Pulsky, and by the time the ship reached South America, he had formally proposed marriage. In the film we are offered a dramatically cogent interpretation of Nijinsky's extraordinary actions: a radiogram is misconstrued, Nijinsky feels himself abandoned, and Romola is there to give solace and support. In the absence of any other evidence, it is as good a key as any to the dancer's precipitate behavior.

We are left with the historical denouement. On September 6 the S.S. Avon docked in Buenos Aires; four days later Vaslav and Romola became man and wife.

It comes as no surprise that Diaghilev took the news badly. Suddenly and without warning, his lover and protégé had been snatched away. Heartbreak soon turned to rage. A telegram was sent to the dancer informing him that his services with the Ballets Russes would no longer be required.

The film draws to a close at this point, but not before indulging in an effective bit of dramatic license. A confrontation between Romola and Diaghilev takes place in his hotel suite. She offers to give up Vaslav for the sake of his happiness and career. The offer is rejected. Such a scene never took place. If it had, there is no reason to suppose that it would have played any differently in real life than in the script.

As the end titles come on screen, we see Nijinsky straitjacketed in an asylum cell, mouthing mystical nonsequiturs as he stares into the camera with vacant eyes. The implication is clear that his abrupt marriage to Romola and Diaghilev's rejection were directly responsible for pushing Nijinsky across the borderline of insanity. Perhaps they were, but the process took rather longer than in the film. To round out the historical picture, we must briefly summarize what life held in store for Nijinsky after the film's foreshortened conclusion.

When it became general knowledge that Nijinsky was no longer associated with the Ballets Russes, a multitude of propositions came his way. None seemed to satisfy him. In the end he foolishly agreed to mount an eight-week season of ballet at a London variety theater. He was to assemble his own troupe of dancers and choose the repertory. But running a ballet company was not Nijinsky's forte. From the start, the enterprise was plagued with mishaps, and when Nijinsky fell ill midway in the second week and was unable to appear, the theater seized on an escape clause in their contract and canceled the engagement. The Nijinskys returned to Budapest to await the birth of their child.* There they were overtaken by the outbreak of World War I.

*She was named Kyra. Twenty-two years later, in a twist of fate almost too fitting to be true, she married Diaghilev's very last protégé, the conductor Igor Markevitch.

As a Russian subject, Nijinsky was technically a prisoner of war, though in the permissive Austro-Hungarian Empire this did not amount to much inconvenience. Meanwhile circumstances were pushing Diaghilev and Nijinsky toward a reconciliation. The war had seriously affected Diaghilev's fortunes. To keep the Ballets Russes employed, he signed a contract to bring the company to America. The engagement was made dependent on Nijinsky's participation. By dint of much high-level diplomacy, the dancer and his family were released from internment in Austria, and on April 4, 1916, they arrived in New York. Diaghilev was at the pier to meet them. "He bowed very low," Romola recounts in her biography of Nijinsky, "and, kissing my hand, offered me a beautiful bouquet. Vaslav, carrying Kyra, followed me closely. Sergei Pavlovitch kissed him on both cheeks, according to the Russian custom, and Vaslav, with a quick gesture, placed Kyra in his arms."

Almost immediately there were disagreements —mostly over money, and most instigated, it would appear, by Romola. To his old colleagues, Nijinsky seemed more withdrawn and reserved than ever, but on stage his artistry and virtuosity seemed unimpaired. Carl Van Vechten, who had followed the dancer's work in Europe before the war, even detected a marked improvement. "I had called Nijinsky's dancing perfection in years gone by, because it so far surpassed that of his nearest rival; now he had surpassed himself."

Another American tour was planned for the autumn and winter. Nijinsky remained in America with Romola and Kyra to await the company's return from Europe. This time the Ballets Russes came without Diaghilev. To avoid friction, Nijinsky was to act as artistic director. The opening weeks at the Metropolitan Opera House saw the creation of Nijinsky's fourth and final ballet, <u>Tyl Eulenspiegel</u>,

Nijinsky in <u>Tyl Eulenspeigel</u>, 1916

of which a few photos survive to tease our imagination. There followed four months of travel across the length and breadth of the United States, with inexorable one-night stands en route—Worcester, Atlantic City, Knoxville, Tulsa, Tacoma. It cannot have been much fun.

In June 1917, Nijinsky rejoined Diaghilev and the company for a series of performances in Spain. Again disagreements arose, this time over the company's forthcoming tour of South America, which Nijinsky was reluctant to take. Diaghilev brought legal pressure to bear, and Nijinsky was obliged to capitulate. It is doubtful whether Romola did much to smooth ruffled feathers. Though he was reluctant to go, Nijinsky performed with his usual zest and brilliance throughout this South American tour. It ended on September 26, 1917, in Buenos Aires, when Nijinsky danced for the last time with the Ballets Russes, appearing in Le Spectre and Petrouchka.

Incipient signs of mental instability are said to have surfaced in South America, but the evidence is contradictory. A year later they were unmistakable. The Nijinskys had settled in St. Moritz to await the end of the war. Completely isolated from the ballet, without any regular routine of practice and rehearsal to impose a structure on his life, unable to communicate with Romola in his native tongue, Nijinsky began to sink into a state of profound depression. Worse was to come. He would parade through the streets of St. Moritz wearing a large golden cross, stopping passers-by to enquire if they had been to mass. Early in 1919, Nijinsky agreed to give a private dance recital for friends and neighbors. It turned out to be a terrifying spectacle. "We felt," Romola wrote, "that Vaslav was like…a tiger let out from the jungle who in any moment could destroy us." Shortly after, he was examined by the distinguished Swiss psychiatrist Eugen Bleuler. He diagnosed Nijinsky as an incurable schizophrenic. As the world's foremost authority on schizophrenia (he coined the term in 1908), Bleuler was in a position to know.

Nijinsky was then 31. He had another 31 years to live—in and out of asylums, sometimes better, sometimes worse, never wholly restored to health. Romola rose magnificently to the occasion. At the beginning she had been attracted to a fantasy—Nijinsky as the embodiment of ballet—and had pursued her fixation relentlessly, unmindful of the consequences. But when better turned to worse, and sickness took the place of health, Romola's love and devotion never wavered. In the final scene of the film, Diaghilev says to her: "I have a suspicion, my dear, that you are by far the best thing that could have happened to him." In the light of history, he suspected rightly.

Over the years various attempts were made to bring Nijinsky out of his mental darkness. Romola moved him for a while to Paris, in the hope that propinquity with old associations might help. In 1928 he sat in Diaghilev's box at the Opéra for a performance by the Ballets Russes. A photograph shows him on stage during an intermission, shyly smiling at Karsavina in her costume for Petrouchka. Count Harry Kessler saw Nijinsky leave the opera house:

> His face, so often radiant as a young god's, for thousands an imperishable memory, was now gray, hung slackly, and void of expression, only fleetly lit by a vacuous smile….Diaghilev had hold of him under one arm and, to go down the three flights of stairs, asked me to support him under the other because Nijinsky, who formerly seemed able to leap over rooftops, now feels his way, uncertainly, anxiously, from step to step.

Later on, insulin shock therapy had some encouraging effect. The dancer Serge Lifar visited him in 1939:

His face had lost its hopelessly timid and down-trodden expression, and he would readily respond to a question or a command....But gone was his sly and childlike smile, a smile of confidence and good-natured benevolence. Its place had been taken by a hoarse laugh, deep and convulsive, which shook his whole body and threw it into sharp and angular plastic poses.

Then another war broke out. The Nijinskys found themselves once again in Hungary—no longer part of the easygoing Austro-Hungarian Empire but a subject state of Hitler's Festung Europa. When the victorious Russian army arrived in March 1945, Vaslav and Romola were hiding out in a village near the Austrian border. Far from upsetting Nijinsky, the presence of Russian-speaking troops drew him out. "It seemed," Romola thought, "almost as if he were awakening from a long and deep sleep." Later that year a correspond-ent for Life found them living at the Hotel Sacher in Vienna. "He is amazingly strong and supple for a man his age," the magazine reported. "Only his face betrays the ruin within, a face...seamed by the fears and suffering of his disease." In 1947, Romola brought him to England, and there he died—of an unsuspected kidney condition—on April 8, 1950.

Midway through the nightfall of Vaslav's insan-ity, Romola contracted to write a biography of her husband in the hope of raising money for his care and treatment. Two expert and dedicated amanu-enses came to her aid: Lincoln Kirstein, then a dance devotee in his mid-twenties, today a veteran critic and animator of ballet in America; and Arnold Haskell, an equally zealous British aficionado, whose book Balletomania kept the legend of

Diaghilev's Ballets Russes alive for a whole new generation of ballet enthusiasts. Published in 1933, Romola's biography quickly became an interna-tional best seller. Although it has since been superseded by more erudite studies—notably the exhaustively researched biography by Richard Buckle (Nijinsky, first published in 1971) and Lincoln Kirstein's annotated compilation of contemporary photographs (Nijinsky Dancing, 1975)—Romola's account of her years with Vaslav remains a document of inestimable interest and value. Nowhere else do we find quite so vivid and compelling a picture of Nijinsky as man and artist.

Its potential as a cinematic "property" was immediately apparent. Only later, after much abortive effort and squandered money, would its inherent difficulties begin to be perceived. Over a period of more than 40 years, the idea of portray-ing Nijinsky's life on screen remained an expen-sive and frustrating chimera.

Vaslav and Romola Nijinsky, 1945

The first attempt came from the producer Alexander Korda, a childhood friend of Romola's in Budapest who had settled in London in 1931. In 1936, Korda proposed making a movie, drawn from Romola's book, in which the actor John Gielgud would be cast as Nijinsky. Before any real progress could be made, the onset of World War II derailed the project, and Korda's rights to the book expired. They were then purchased by an American theatrical producer, Oscar Serlin, who had recently brought the long-running dramatization of Clarence Day's Life with Father to Broadway. Serlin believed that Nijinsky's life story held similar promise, and he discussed the possibility of a dramatization with several writers, among them Clifford Odets. These explorations yielded nothing that was satisfactory, and as the years dragged on Serlin lost interest. Romola came to see him soon after Nijinsky's death in 1950—without, it would seem, making much headway in reviving his enthusiasm.

A few years later, she sold the rights to another Hungarian emigré film director, Charles Vidor. Vidor had come to the United States as a Wagnerian singer and soon gravitated to Hollywood, where he made a specialty of directing tearful romantic dramas with musical overtones. One of these was A Song to Remember, in which Cornel Wilde went through the uncertain motions of impersonating Chopin; another was Rhapsody, in which a wealthy woman (Elizabeth Taylor) becomes emotionally entangled with a violinist (Vittorio Gassman) and a pianist (John Ericson). Given his penchant for undertaking films with vaguely cultural aspirations, it is not surprising that Charles Vidor lent Romola a willing ear. In 1954, the year of Rhapsody, he bought the rights to her biography of Nijinsky. A long lawsuit ensued, brought by Serlin, who claimed that the rights still belonged to him. Serlin eventually lost out over a technicality; but while the case was being argued, it seemed only prudent to hold Nijinsky in limbo. The project was still on the shelf when Charles Vidor died of a sudden heart attack in 1959. Considering his predilection for the mawkish and the banal, it is probably just as well that Nijinsky escaped his attentions.

The curtain now descends for a decade. When it rises again, in 1969, the Western world had acquired a new dance idol from Russia in the person of Rudolf Nureyev. Hailed everywhere as the successor to Nijinsky, Nureyev appeared to be cast by heaven for the starring role in a film about his legendary antecedent. So it seemed to an independent film producer, born in Canada but resident in London, named Harry Saltzman. On the face of it, Saltzman was an unlikely candidate to take up the cause of Nijinsky. His forte as a producer had manifested itself mostly in glossy thrillers, such as The Ipcress File and the early James Bond films. But unlikely or not, it was Saltzman who got in touch with the widow of Charles Vidor and purchased from her the languishing film rights to Romola's biography.

With Nureyev under option to play Nijinsky, Saltzman set out to get the long-awaited film into production. It would be fruitless to chronicle his myriad setbacks in detail. Few film projects can have abraded so many tempers or confounded so many clever professionals. It consumed the time and expertise of directors as diverse as Ken Russell, Jerome Robbins, Tony Richardson, and Franco Zeffirelli. It engaged the participation of writers with such high credentials as Harold Pinter, Edward Bond, Melvyn Bragg, and Edward Albee. But despite—or because of—this profligate pooling of creative talent, there was never a meeting of minds. Producers and writers came and went, large sums were spent on travel, conferences, research, preliminary drafts—all to little effect. After several years of costly false starts, Saltzman had had enough. He sold an option on the Nijinsky rights to a fellow producer, David Picker, onetime head of United Artists. But Picker was no more successful than his predecessors in achieving a workable collaboration. In 1977 his option ran out and the rights to Nijinsky's life story reverted to Saltzman.

That year a film opened which was to cast a new light on the commercial potential of ballet in movie theaters. The Turning Point was unequivocally centered around the world of dance. To be sure, it had a good dramatic story line, but the plot unfolded almost entirely against a background of ballet classes, rehearsals, and performances. It employed famous dancers in key roles—Mikhail Baryshnikov, Antoinette Sibley, Alexandra Danilova. And it was a huge box-office success. By good luck or good timing, the film coincided with a remarkable upsurge of interest in ballet all over America. In the preceding decade, the audience for ballet had increased from a million people, mostly centered in and around New York, to some sixteen million throughout the country. New ballet companies had put down roots in every major American city, with year-round local seasons to supplement the regular tours by such bellwether troupes as American Ballet Theater and the New York City Ballet. Newspapers and magazines marveled in print about "the dance explosion." At the very crest of this floodtide, The Turning Point came along to tap and nurture ballet's burgeoning constituency. Enter Herbert Ross, the film's director, and his wife Nora Kaye, its prime instigator.

Tall, curly-haired, bespectacled, given to wearing plaid sport shirts and baggy trousers, urbanely low-key in speech and manner, Herbert Ross could more easily be taken for a college professor than a big-money movie director. Appearances are deceiving. He has been in show business all his life. Born in Brooklyn, Ross started out as a teen-age actor but soon set his sights on a career in the ballet. "I was," he confesses, "in some ways ill suited for it. I was tall and didn't have the natural attributes, but I studied like a maniac." To pay for ballet lessons, he worked as a hoofer in Broadway shows. Sidelined with a broken ankle during the run of Inside U.S.A., he choreographed his first ballet, Caprichos, in 1950. By this time the infant television industry was gobbling up promising talent right and left. Ross was among those inducted, working first as a choreographer, then as a director, and later as a producer on "The Milton Berle Show," "The Martha Raye Show," and "The Bell Telephone Hour." In his free moments he served as choreographer on several Broadway musicals.

The pace eventually wore him out. In 1958, Ross quit television and returned to ballet. Within a year he had created three new works for American Ballet Theater and married the company's celebrated prima ballerina, Nora Kaye. Together they formed their own company to tour through

Herbert Ross and Nora Kaye

Europe from headquarters in Brussels. It lasted only a couple of years. "I hated it," Ross says. "It was what I thought I had always wanted, but I didn't want it at all." When the company dissolved, Nora Kaye announced her retirement after 22 years on stage. She was to find a new career in Hollywood.

Back in America, Ross became a much-sought-after collaborator on screen musicals. Starting as a choreographer working under other directors' supervision, he soon began directing entire musical sequences on his own, most notably those for Barbra Streisand in Funny Girl. This in turn led to his first film as full director, the musical version of Goodbye, Mr. Chips, in 1968. Thereafter, with Nora Kaye's increasing collaboration, he embarked on a series of pictures that won wide recognition for high craftsmanship and good taste, among them The Seven-Per-Cent Solution, Play It Again, Sam, and The Sunshine Boys. By the time he won an Academy nomination for The Turning Point, Ross already had eleven films to his credit.

Harry Saltzman, buoyed by the immediate and unexpected success of The Turning Point, saw in Ross the possible savior of his stillborn Nijinsky project. They met shortly after the film's New York opening. Ross's first reaction was tepid. "I had been hearing about the Nijinsky film for years, and it seemed to me like a very shopworn proposition. So when Harry brought it up, it was as if he were asking me to do Remembrance of Things Past or The Bible—one of those hopeless projects that never gets off the ground. I also didn't have a clue as to how it could be done. But I guess I really wanted to get involved." Before long, he and Saltzman had come to an understanding. Saltzman was to subsidize the first draft of a new film script; if it met with Ross's approval, he would then make a deal with a major production company to finance the film. From the start it was agreed that Ross was to exercise full control.

The first task was to choose a writer. "Who was left?" Ross asks rhetorically, and only half-jokingly. "So many people had worked on the script, and failed, that relatively few writers came to mind who were equipped either in interest or attitude to handle the material. But from the start I had a hunch that Hugh Wheeler could do it." The two met in New York toward the end of 1977. By then, Ross knew what he didn't want. "All the other scripts, I gathered, had taken off from Nijinsky's insanity. They showed a madman reliving his triumphs in flashback. I felt that was a fatal mistake. If you begin with the insanity, you rob the story of all drama. Hugh was in full agreement. We both saw it as a film about complex and fascinating human relationships, not a study of schizophrenia."

Hugh Wheeler, born in London but an American by long residence and citizenship, came to films late in his career. For the first twenty-five years of his writing life, he ground out mystery novels under various pseudonyms (many of them with Richard Webb as co-author). His first play, Big Fish, Little Fish, opened on Broadway in 1961. Though it had only a moderate run at the time, the play has since developed a cult reputation as a minor American classic. Over the next decade Wheeler's work was uneventful: two plays, both of them adaptations and both failures, and two scripts for forgotten movies. Then in 1973, with A Little Night Music, he began a series of collaborations with Stephen Sondheim and Harold Prince that finally brought him, in his early sixties, the balm of commercial success. As Wheeler's stock rose on Broadway, so did his reputation in Hollywood. He has been responsible in recent years for the screenplays of Cabaret and Travels with My Aunt. Herbert Ross's choice of this late bloomer for the thorny Nijinsky assignment was thus not altogether fortuitous. Hugh Wheeler was a writer of long experience and proven capability.

Nevertheless, Wheeler's first draft was disappointing. "I had read all the available material," he recalls, "and in the first draft I tried—as one usually does in the early stages—to put everything in, because everything was potentially so interesting." But despite the prolixities and sometimes discordant tone of this initial script, there was one scene that impressed Ross greatly—a scene, at the very start of the screenplay, between Diaghilev and Nijinsky. "I had never read a scene like it before," Ross says. "It was so matter of fact, so candid, so honest. It depicted a relationship between two men without apologies, without snickers, without pussyfooting. I liked the concept, and I saw the seed there for a really remarkable film script." Wheeler adds: "The scene indicated that it was the intimate personal relationship between the two men—rather than unending shots of triumphant performances, glamorous parties, and the like—which could most effectively show how a work of art comes into being. Art is a very mysterious thing, and so is a genuine—however offbeat—relationship of love between two individuals. It seems to me (and to Herbert Ross, whose instincts were at all times relatively congruous with mine) that even the greatest talents are fragile beyond words and that it is only in the exactly right climate, however improbable, that they can flourish."

During the early months of 1978, the screenplay for Nijinsky went through a number of successive drafts, as Wheeler—working closely with Ross—cut, shaped, refined, and polished. By May, Ross was satisfied and ready to go ahead. But before we reach the next step in the film's gestation, it would be well to pause for a brief look at Hugh Wheeler's final shooting script.

The screenplay can conveniently be divided into five chief locations—Budapest, Monte Carlo, Paris, the S.S. Avon, and Buenos Aires—separated by interludes in Greece, Berlin, London, and Venice, and ending with a coda in Genoa. (Several of these locations, it should be noted, are not explicitly identified in the movie, which appears less fragmented geographically than the above itinerary might indicate.) In Budapest, over a period of roughly 24 hours, we meet most of the major protagonists: Nijinsky, on and off stage; Diaghilev, just back from Paris and London with

new contracts; Romola, getting her first glimpse of the Ballets Russes; Karsavina, arriving from St. Petersburg to dance with Nijinsky; Fokine, already uneasy at the prospect of a rival choreographer in the company. We are also introduced to important subsidiary characters: the ballet master Cecchetti, the business manager Astruc, and the servant Vassili. As the action moves from hotel to railroad station to rehearsal room to theater lobby, we learn something of the camaraderie, the glamour, and the hard work that make up the daily routine of a touring ballet company. And before the camera leaves Budapest, we are left in no doubt about the intimate and fruitful relationship between Diaghilev and Nijinsky (the scene that so impressed Herbert Ross in the first draft), nor about the romantic fantasies of Romola de Pulsky.

An interlude in Greece, where Diaghilev takes Nijinsky to stimulate the creative juices for L'Après-midi d'un Faune, leads to Monte Carlo and Romola's arrival there in pursuit of the Ballets Russes and its leading dancer. (Wheeler describes her as "a crypto-proto-groupie.") Here we meet the Baron de Gunzburg and are shown an example of his shirt-cuff accounting. It is through De Gunzburg's mediation that Romola meets Diaghilev at a grand reception and learns that she might possibly be allowed to join the company as a paying pupil of Cecchetti. Meanwhile, Fokine's resentment reaches boiling point, and there is an angry confrontation in Diaghilev's Monte Carlo hotel suite.

A bridging scene in a railroad car takes us to

Paris for the scandal-making first performance of <u>Faune</u>. Fokine's definitive break with Diaghilev is sealed at this time. Another interlude—this one in Berlin—shows Nijinsky rehearsing <u>Le Sacre du Printemps</u> amid growing tension and frayed tempers. The script has Stravinsky complaining to Diaghilev: "The boy's hopeless. He hardly understands music at all. He has no ear. He can't even count." Stravinsky did indeed voice sentiments along these lines in his 1935 autobiography, but Robert Craft has recently published evidence* that the composer and choreographer were in fact very much in accord throughout the rehearsals. At the end of his life, Stravinsky recanted the animadversions in his autobiography and reaffirmed his original approval of Nijinsky's contribution.

From the <u>Sacre</u> rehearsals the film script takes us back to Paris for a scene between the Baron and Nijinsky. This leads into the first performance of <u>Jeux</u> (May 1913) and to the dancer-choreographer's backstage disappointment at its unenthusiastic reception.

An interlude in England ensues. At Lady Ripon's country mansion, just outside London, one of the film's opulent party scenes is in full swing. While Nijinsky entertains the guests, Diaghilev learns the shattering news that the Narodny Dom in St. Petersburg has just burned to the ground. An alternative engagement for the autumn will have to be found immediately. The Baron suggests a tour of South America. Here the screen writer fiddles a bit with the facts. As we have noted earlier, the Narodny Dom fire actually took place more than a year before, and was responsible for bringing the company to Budapest in 1912—and hence ultimately for bringing Vaslav into Romola's orbit. But Wheeler's poetic license is altogether defensible. By making the fire responsible for the

*<u>Stravinsky in Pictures and Documents</u> (New York, 1978).

South American tour, he ensures that its effect on the Romola-Vaslav relationship is every bit as immediate.

Again the script returns us to Paris, where we witness the riotous first performance of <u>Le Sacre du Printemps.</u> The morning after, Astruc reads aloud an uncomplimentary newspaper review and questions whether Diaghilev should still entrust the new Richard Strauss ballet, planned for the next season, to Nijinsky. For the first time, Diaghilev fails to spring to his protégé's defense.

Another interlude now brings us to an unspecified beach resort (though in fact Diaghilev and Nijinsky spent their last holiday together in Baden-Baden). It is here that Hugh Wheeler attempts to give dramatic articulation to the malaise affecting their relationship. In his view, the seeds of

self-destruction were always present. Let him explain:

Diaghilev was a genius with no creative talent of his own. He needed an instrument through which he could express his otherwise incommunicable artistic energy. Nijinsky, who fulfilled that role, obsessively needed a stronger, dominant, loving figure (a father figure, if you like) as a shield from a world he was almost entirely incapable of coping with. Two exceptional people had found exactly the right balance for themselves. It was tragic but inevitable that the negative aspects of the relationship would eventually destroy it. Nijinsky was split between his great need for Diaghilev and the natural youthful desire to "find his independence." A neurotic, self-downgrading anxiety started him believing (incorrectly) that Diaghilev was getting tired of him and would throw him out.

The manifestation of that neurosis is dramatized in this last scene between the dancer and his mentor.

The S.S. Avon is now at sea, carrying the entire company to South America, except for Diaghilev. At last, Romola has an opportunity to stalk her prey. Throughout the film she has been endeavoring ineffectually to make Vaslav acknowledge her existence. Now, adrift from Diaghilev for the first time in years and seized with doubts about the status of their relationship, he begins to pay attention to her. Nijinsky's anxiety is heightened by a wireless message to Baron de Gunzburg, which suggests that Diaghilev is engaging another dancer for the Strauss ballet and is entrusting its choreography to Fokine. The message is Wheeler's invention, but it serves admirably to motivate and explain Nijinsky's gathering panic. Like a ship caught in a sudden storm, he heads for the nearest port—which happens to be the sympathetic and supportive Romola. But why, out of the blue, did he—an avowed homosexual for the past five years

or so—propose marriage after an acquaintanceship of only a few days? As Wheeler explains it,

Nijinsky's ego—very frail at best—made him want to believe that he could stand on his own. This, coupled with the accident of Romola's fierce determination to get him, brought him to plunge into the marriage. A marriage, he thought (if he thought about it at all), would assert not only his own manhood but would throw a challenge to Diaghilev, who—in Nijinsky's fantasy—would instantly do his utmost to win him back.

All this is expressed dramatically in a series of compelling scenes on board ship. Sticklers for detail may object that Vaslav and Romola could not possibly have communicated with each other so volubly, since they had at the time no common language (the Baron, in actual fact, served as interpreter for Vaslav's proposal of marriage). But there is emotional if not literal truth in the shipboard encounters that Wheeler has imagined.

In Buenos Aires, the last principal location, Vaslav and Romola are joined in marriage. Imme-diately thereafter Karsavina arrives, bringing the Strauss score with a message from Diaghilev that Nijinsky is to dance the main role as well as create the choreography. So the anxiety on board ship was wholly without foundation. Now Vaslav is seized with even greater panic as he speculates on Diaghilev's reaction to the news of his marriage. It comes soon enough in the form of a cable from Astruc, informing the dancer that his services with the company will be permanently termi-nated at the end of the South American tour.

Vaslav swings uncontrollably from violent hys-teria to apathetic withdrawal. All this considerably anticipates events. The telegram severing Nijin-sky's contract was not sent until well after the South American tour had ended, and there is no evidence that Nijinsky showed signs of mental instability at this time, but the script is building to a climax, and the essentials are accurate if not the chronology.

The film's coda takes place in Genoa, where the company disembarks from its South American tour. Nijinsky still naively believes that Diaghilev will be at the gangplank to greet him. The shock at not finding him there is devastating and destructive.

Diaghilev [Wheeler contends] was the victim of his own pride and his own overdeveloped "worldly wisdom." This made him believe that Nijinsky under Romola's influence would become far too complicated to serve as a malleable interpreter of Diaghilev's own artistic impulses, and it caused him to act in a manner diametrically opposed to Nijinsky's expectations. It was a disaster for both of them—and also for Romola, who was in no way a villain, but only a tremendously strong-willed, naive young girl who had fallen in love with a fantasy and assumed, with total honesty, that she could turn the fantasy into a husband.

Once in Genoa there remains only the confrontation between Romola and Diaghilev, a wholly imaginary encounter to which we have already alluded, before the final end-titles superimposed on a straitjacketed Nijinsky.

This skeletal précis of the screenplay does not begin to do justice to the texture of Wheeler's characterizations or the charm of his writing, but it will at least serve as orientation when we come to examine the making of the film. That process is still some time off. Before the camera begins to turn, much groundwork remains to be accomplished.

Ross's first priority, once Paramount had agreed to finance the film, was to cast the role of Nijinsky. Time had by now eliminated Rudolf Nureyev as a candidate. In 1978, at the age of 40, he could no longer qualify for the part of a young dancer in his early twenties. Mikhail Baryshnikov, another and younger emigré from Russia, was a more likely contender. Ross had given Baryshnikov an important part in The Turning Point—that of the dancer Yuri, who performs breathtaking feats of virtuosity on stage and is the object of every ballerina's and would-be ballerina's affections. In the very early stages of their involvement in the Nijinsky film, Ross and his wife had more or less assumed that Baryshnikov would be cast in the title role. Certainly, they were never in any doubt that he could convincingly re-create Nijinsky's prowess as a dancer. (Nora Kaye believes, indeed, that the standard of dance technique has advanced so tremendously since Nijinsky's time that dancers on the level of Baryshnikov and Nureyev have far surpassed him, at least in terms of sheer virtuosity.) Nevertheless, Baryshnikov too was eliminated from consideration for the Nijinsky role. Aside from the problem of his heavy accent (in The Turning Point he had relatively few lines), the Rosses reluctantly concluded that Baryshnikov at age 30 was already too mature for the part, and—even more significantly—had become too assured and established as a personality in his own right.

"We felt," Ross explains, "that Vaslav and

Romola had to be played by very young and naive and unsophisticated people. They had to be guileless—two children, really, who opened up this Pandora's box of emotion and didn't know how to close it. One always has the tendency to envisage historical figures as being older than they really were. Nijinsky had just turned 24 when Romola first saw him dance, and he was a very immature 24 at that. When Diaghilev calls him 'a neurotic child' in the script, he isn't exaggerating.

So we had to find someone in his early twenties who not only resembled Nijinsky in physical appearance but also had the same quality of immaturity and vulnerability and androgyny, someone who could dance well enough to sustain the illusion of being the sensational star of the Ballets Russes, and someone who could act an extremely demanding dramatic rôle. You could say that our options were not exactly limitless."

Only two candidates were ever considered, and one of these fell out of the running early in the race. This was Patrick Dupond, a young French dancer who had come to worldwide attention in 1976 by winning the Varna International Competition. The Rosses had seen him dance in Chicago and were much impressed. They were also struck by his resemblance to Nijinsky. But when the time came to make a screen test, it became evident that Dupond's shaky command of English would pose too great a hurdle.

That left George de la Pena. Born in New York in 1955 of mixed Russian-Argentinian parentage, he fell into a dancing career entirely by accident. His mother, from whom he learned Russian even before he began to speak English, was an accom-

plished pianist, and it was always assumed that George would follow in her footsteps. But when he applied for admission to New York City's High School of Performing Arts at the age of thirteen, there were no openings in the music department. He signed up, _faute de mieux_, for ballet instead. "The teachers started saying I had potential," he says, "though I suspect it was only because they needed male dancers. Ballet had never been one of my ambitions, but when I saw Nureyev dance _Swan Lake_ with the Royal Ballet, my attitude changed. I couldn't believe that dancing could be so athletic and at the same time so beautiful."

After graduating from the high school, he was accepted as a trainee at American Ballet Theater, and soon after began appearing in the company's corps de ballet.

The Rosses became aware of him almost immediately. Nora is an associate director of American Ballet Theater and as such keeps a close watch on the company's recruits. In The Turning Point they engaged him to serve as Baryshnikov's stand-in. By that time De la Pena was already dancing principal roles, and inevitably he came to mind for the Nijinsky assignment. He was the right age, he had the right looks, and he could dance. The big question mark concerned his acting ability. From observing him on stage, Ross was satisfied that the dancer could act persuasively within the context of ballet. But acting before the camera poses a wholly different kind of challenge. "The problem for dancers," Ross says, "is speech, because they have trained their instrument—the body—to express everything and have totally neglected the culture of the voice."

In New York, Ross listened as George de la Pena read aloud some key scenes from the script. The results were sufficiently encouraging to take him to the West Coast for a more rigorous tryout. There, properly made up and lit, the dancer made a four-minute screen test that left Ross in no doubt about his ability to project via the camera lens. The test was shot on August 23, 1978. Paramount's top executives saw it a few days later and gave their benediction. For better or worse, the part of Nijinsky was to be played by an unknown.

The tempo of events now began to accelerate. A decision had already been made to base the production in England, and there the foundations were being laid by Stanley O'Toole, a British film executive who had previously worked with Ross on The Seven-Per-Cent Solution and The Last of Sheila.

On Nijinsky he was sharing production responsibility with Nora Kaye. While she and Herbert were still in America, O'Toole set up headquarters at Pinewood Studios, just outside London, and began assembling the nuts and bolts of a highly complex production. A large crew had to be hired, legal and insurance matters put in order, and arrangements set in motion for filming on location all over Europe. Every move, moreover, had to be carefully assessed in terms of the production's overall budget. These preliminaries were well in hand when Nora Kaye arrived in mid-September to begin coping with the sets, costumes, and staging of the ballet sequences. Ross joined her in London on October 1. In less than four months, shooting was scheduled to start.

"The hardest part of any picture for me," Ross says, "is the pre-production planning. Every decision you make affects the film's destiny. Nijinsky was especially hair-raising because it was so complicated and we had so little time. I got to it late because of some last-minute work on California Suite [his previous film]. So we had just about three months to fit all the pieces together. Every day brought a fresh crisis. There were endless lists, endless research, endless auditions, endless travel, endless conferences."

Casting was of foremost concern. When Ross arrived in London, only the role of Nijinsky was definitely assigned, though by then there had been auditions in America for some of the other parts. In the early days of the project he had invited Leslie Browne to test for—or, in the jargon of the theater world, "read"—the role of Romola. The young dancer-actress had just won an Academy Award nomination for Best Supporting Actress as the fledgling ballerina in The Turning Point, and Ross considered her a prime contender for Romola. Nevertheless, he made a wide search for other candidates. "I read a number of actresses in New York and California, and then when I came to London I must have read at least twenty more. The role demands above all an ability to project an innocent obsession. Truffaut's The Story of Adèle H had dealt with that kind of obsession, and the thought crossed my mind that Isabelle Adjani might be right for the part. But she was tied up making a film about the Brontë sisters, and anyway I'm not convinced that it's possible to play a complex part in a language which is not your own. So after exhausting all the possibilities, I came back to Leslie. None of the others seemed to have

Leslie's peculiar quality—that strange mixture of total naiveté and innocence combined with a latent carnality."

Guileless innocence and rudderless youth might be all very well for Romola and Vaslav, but for the character of Diaghilev the requisites were altogether different. "It's a very dangerous part," Ross declares. "I wanted at all costs to avoid giving the impression that Diaghilev was some sort of monster or that he was in any way depraved or distasteful. So I thought it essential for the part to be played by a physically attractive man, someone still young enough to be a credible partner in a relationship with a very immature boy. Again, we have a tendency to think of Diaghilev as being much older than he was. In fact, he was just turning 40 when the film opens. So I wanted an actor of about that age who would seem virile and romantic and appealing. He had to be able to convey sophistication and culture and wisdom, and to project a concern for art without seeming overly esthetic or overly refined. And he had to be utterly charming. We were exceedingly fortunate to find all those qualities in Alan Bates."

At their first meeting, Bates seemed a little hesitant about taking the role. Ross thought that the prospect of appearing as a notorious homosexual might have been somewhat daunting to an actor fresh from his success as a latter-day Prince Charming in An Unmarried Woman. "Actually, it wasn't that at all," Bates says. "It was simply that I had just been offered two other very good scripts, and I needed a little time to decide. In the end, of course, I had to play Diaghilev. He was such an extraordinary man. And it was a new experience for me to portray a real historical personage. I could actually talk to people who had known and worked with Diaghilev, and that made the role much more exciting and challenging. Fortunately, it

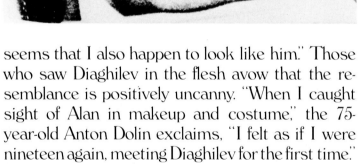

seems that I also happen to look like him." Those who saw Diaghilev in the flesh avow that the resemblance is positively uncanny. "When I caught sight of Alan in makeup and costume," the 75-year-old Anton Dolin exclaims, "I felt as if I were nineteen again, meeting Diaghilev for the first time."

English by birth, training, and residence, Bates moves readily—like so many other members of the British theatrical establishment—between stage, film, and television, and between the West End, Broadway, and Hollywood. He first came to attention, at the age of 22, as a member of the original cast of John Osborne's Look Back in Anger (1956), and since then has appeared in plays by many of England's leading contemporary playwrights. His notable films include The Entertainer, A Kind of Loving, Georgy Girl, Women in Love, and The Go-Between. Bates guards his privacy closely and is not much given to oracular pronouncements about his career or his style as an actor. Of his approach to Diaghilev, the most he will say is that he read all he could about the man, talked to those who had known him, and then played what was written in the script as best he could.

In choosing the rest of the cast, Ross drew largely on London's vast pool of acting talent. Auditions began immediately after his arrival there early in October. Some of the actors and actresses he knew already, others were suggested by his casting adviser, Rose Tobias Shaw. Alan Badel, a distinguished veteran, was an easy choice as the charming, if feckless, Baron de Gunzburg. Ronald Pickup, long associated with the National Theatre, seemed ideally suited both in looks and background (his second love is music) for the role of Stravinsky. And Jeremy Irons, a young player with the Royal Shakespeare Company, fitted Nora Kaye's exacting requirements for the role of Fokine (she had known the real Fokine as a child student years before). Even minor parts were filled by highly accomplished performers. The role of Romola's mother, for instance, went to Janet Suzman, an acclaimed Beatrice, Rosalind, and Cleopatra at the Shakespeare Memorial Theatre in Stratford-on-Avon.

Two important supporting roles were cast from the world of ballet. It was Nora Kaye's idea to assign the part of Enrico Cecchetti to Anton Dolin, who brought to the production a precious living link with Diaghilev and his world. Dolin, born Patrick Healey-Kay, had joined the Ballets Russes in 1923 at the age of nineteen, and appeared with the company off and on as a leading dancer until Diaghilev's death in 1929. Alone among the members of the film's cast, Dolin had known Diaghilev and Romola, had even visited the semi-invalid Nijinsky in Paris during the 1920s. He had also known the character he was portraying, though his memories of the exigent ballet master are not exactly haloed with kindly feelings. "Cecchetti," he recalls, "was rather morose and tyrannical. We had a morning class with him every day except Sunday, and I hated it. My three teachers had all been Russian, and I was totally committed to the Russian approach. Cecchetti's methods seemed alien to me, and I rebelled. Doubtless I exaggerated the differences." In her memoirs, Karsavina tells how Cecchetti would playfully lash out at a tardy dancer with his walking stick. When Dolin recreates this little piece of foolery in the film, he wields a cane that actually belonged to Cecchetti.

The role of Karsavina herself remained unfilled until a few weeks before shooting began. Once again the options were limited. The part had to be played by someone who could not only dance well but also convey the attitude of a glamorous and acclaimed ballerina. "We must have read 40 dancers for the part," Ross says, "including many of the great ones in America and Europe. The role seemed to lapse into caricature. Either it would come out looking like a rip-off of the part Antoinette Sibley played in The Turning Point, or—worse— it would come out like a corny Russian ballerina in a musical of the 1930s. We didn't get to Carla Fracci until the very last moment, and by then Nora and I were really uptight. Fortunately, Carla was absolutely right for the role. She came twice to read the part because I was unsure of her accent. But we decided that any problems with her accent would be far outweighed by the positive qualities she could bring to the role. Nobody but a dancer of her stature and experience could communicate such total assurance." La Fracci, as she is known south of the Alps, is Italy's foremost ballerina. A star at La Scala for many years, she has since appeared as a guest artist throughout Europe and America in the company of such partners as Nureyev, Baryshnikov, and Erik Bruhn.

All this time, while casting was in progress, preparations were being hurried along for staging no less than seven ballets for the film. Scattered through the screenplay were sequences involving excerpts from four works by Fokine (the "Polovtsian Dances" from Prince Igor, Le Spectre de la Rose, Schéhérazade, Petrouchka) and three by Nijinsky (L'Après-midi d'un Faune, Jeux, Le Sacre du Printemps). Each had to be re-created as closely as possible in detail and spirit to the original production. (Much later, when editing the final cut, Ross deleted the Prince Igor sequence in order to tighten the story line.)

Fokine's ballets were the easier to do. Two generations of dancers and balletomanes had been weaned on these staples, which remained in the repertory of the Ballets Russes and its various successor companies until the late 1940s. Then, quite suddenly, the ballets of Fokine were crowded off the stage by newer creations from Balanchine, Tudor, Robbins, Ashton, Cranko, MacMillan, and other contemporaries. By the late 1970s, only one company in the West—the London Festival Ballet —still regularly performed the four Fokine ballets planned for the film. This London-based but mainly peripatetic troupe, formed in 1950 by Anton Dolin and Alicia Markova, had over the years attracted a devoted public in Britain (without, it must be added, ever challenging the artistic stature of the Royal Ballet). During a visit of the London Festival Ballet to New York in July 1978, the Rosses had seen its productions of the Fokine repertory, and they were sufficiently impressed to engage the company for the film.

Ballet tradition is handed on by a succession of torch bearers, the dancers of one generation passing on what they have learned to those of the next. The torch bearer for London Festival Ballet's Fokine revivals was Nicholas Beriozoff, a Russian dancer and (later) ballet master who learned the ballets in the 1930s from veterans of the Diaghilev Ballets Russes. Beriozoff's restagings form the basis of the Fokine ballet sequences seen in the film, though various refinements were introduced as other experts contributed further advice. In pursuit of authenticity the Rosses had brought along their own torch bearer in the person of Irina Baronova, a Russian-born ballerina who had danced in all the Fokine ballets during the 1930s and 1940s. Mme. Baronova was on hand throughout rehearsals and shooting to add the benefit of her own long experience in this repertory.

Another expert engaged for the production was Nicholas Georgiadis, distinguished ballet designer and lecturer at the Slade School of Fine Arts in London, whose responsibility it was to evaluate and—where necessary—revamp the London Festival Ballet's existing sets and costumes. "When we went back to the sources," Nora Kaye relates, "we discovered that a lot of things needed fixing. In Spectre, for instance, we had to repaint the colors of the sets to match those in the Bakst designs. We also made a new chair to conform more accurately to the original, and we had a new rose-petal costume designed for George. In Schéhérazade we ordered new costumes for the principals and had new brass filigree lamps made to match the originals. And in Petrouchka we made a new set for the Cell Scene."

Nora Kaye is confident that the reconstitutions of Fokine's ballets in Nijinsky come closer to the original productions than anything seen for half a century. And she hopes that the film will serve to stimulate a revival of interest in the work of her old teacher and idol. "For years," she says, "people dismissed Fokine's choreography as old-fashioned. Well, probably it is. But I think we should look at the Fokine ballets the way we look

at old movies—as period pieces with their own validity. The trouble is that ballet companies are always wanting to do new things. So it's not surprising that dancers no longer know how to handle the style of Fokine's ballets."

George de la Pena had to be rigorously coached in that style. During the autumn of 1978, while the Rosses coped with pre-production problems in London, the young star of the film worked in New York with three veteran dancers from the Ballets Russes de Monte Carlo of the 1930s: Fredric Franklin, who had himself studied with Nikolay Legat, one of Diaghilev's ballet masters; Yurek Lazowski, a Polish character dancer, famous for his Petrouchka; and Vladimir Dokoudovsky, whose teacher, Olga Preobrazhenska, had appeared in Fokine's early ballets at the Maryinsky Theater. With this newly acquired knowledge of the Fokine roles, George de la Pena has become —whether he realizes it or not—a torch bearer himself, equipped to pass on the tradition to dancers yet unborn.

Nijinsky's ballets proved to be far more elusive. Of the three, only L'Après-midi d'un Faune had survived its initial hostile reception. Diaghilev retained it in the Ballets Russes repertory for many

years. The ballet had even reverted to its original interpreter when Nijinsky briefly returned to the company in 1916-17. During this period, Nijinsky coached one of Diaghilev's new recruits, a Polish dancer named Leon Woizikowski, in the role. Woizikowski danced Faune for the Ballets Russes throughout the 1920s, and in 1931 he staged and danced it for Ballet Rambert in London. "Many dancers appeared in Faune," Nora Kaye says, "but Woizikowski's version was always the closest to Nijinsky's. Unfortunately, we couldn't get Woizikowski to help us, because he died in 1975. So we got the dancer who followed him in the role at Ballet Rambert." This was William Chappell, now in his mid-seventies, from whom De la Pena began learning the role as soon as he arrived in London in early December. With the help of another Rambert veteran, Elizabeth Schooling (who

had danced the First Nymph), Chappell re-choreographed the entire ballet, while Nicholas Georgiadis re-created the sets and costumes from the original drawings by Bakst and the excellent photos of the first Paris production.

For L'Après-midi d'un Faune there was a thin thread of memory leading back to the early history. For Jeux and Le Sacre there was nothing. Both ballets had been unceremoniously dropped from the Ballets Russes repertory after their disastrous premieres in 1913. Jeux never resurfaced. When Diaghilev wanted to revive Le Sacre in 1920, nobody in the company could remember how it had been danced. (The various systems of dance notation in use today had not yet been perfected.) Thus, nothing survives of either ballet except some photographs of the original productions (rather skimpy in the case of Le Sacre) and the drawings of Valentine Gross.

To fill this void, Kenneth MacMillan was commissioned to devise new choreography in the style of Nijinsky. MacMillan is the director of the Royal Ballet and one of the most successful choreographers of our time. His job, he says, was to function

"as a sort of Walt Disney. We have the photographs and the Valentine Gross drawings, and I've really just animated them." There are further hints, of course, in the comments and descriptions of those who witnessed the first performances, notably the long analysis of <u>Le Sacre</u> by Jacques Rivière. But static photos, drawings from memory, and passages of descriptive prose do not begin to convey the complexities of ballet movement, and Kenneth MacMillan would undoubtedly be the first to concede that his re-creations of Nijinsky's two lost ballets are at best inspired guesses. Once again Nicholas Georgiadis, a longtime collaborator on MacMillan ballets, designed sets and costumes according to the photos and drawings of the Paris productions.

According to plan, Georgiadis was to have supervised the design of sets and costumes for the entire production. This did not prove to be practical, and in the end his contribution was limited to the ballet sequences alone. Responsibility for the remainder of the film fell to two London-based specialists, both with wide experience in theater and film: John Blezard for the sets, Alan Barrett for the costumes. They joined the production less than a month before shooting started.

Somehow, in the midst of all the trials and travails in London, Ross managed to get away on two occasions for a reconnaissance trip (or "recce" in the language of filmdom). Traveling by private plane along with Nora and other members of the production team, he made the rounds of various proposed location sites on the Continent. The aim was to attain maximum authenticity with minimum dislocation. "At the beginning of a movie project," he notes, "you're not prepared to compromise on anything. If the script has a scene in Buenos Aires, you think of going to Argentina to film it. Then, bit by bit, you begin to simplify the logistics. <u>Nijinsky</u> is more peripatetic than most films, in that the ballet company was constantly on the move, and obviously we couldn't go to every location mentioned in the script. Budapest, I thought, was essential. The film opens there, and I wanted to convey the city's rather special turn-of-the-century atmosphere. The French Riviera also seemed indispensable, because it evokes so well the world of grand hotels and grand society in which Diaghilev and Nijinsky moved. Another decision, which we had made from the start, was to shoot the Greek temple scenes in Sicily, because the temples there are among the most beautiful and best preserved in the Mediterranean. Sicily turned out to be a bonanza. On our first recce there, we found a number of <u>belle époque</u> villas that were ideal for the party sequences, as well as a good church facade for the wedding scene and a lovely opera house for interior shots." Ross thereupon decided to limit the filming sites (beyond London and environs) to Budapest, the French Riviera, and Sicily. Curiously, nothing in the film script is actually set in Sicily. And, conversely, though much of the action is supposed to take place in Paris, no footage was filmed there. Thus does art triumph over geography.

"You come back from your first recce," Ross says, "with a blur of impressions. Then you try to sort out all the things you have seen, fitting locations to scenes in the script, and sometimes altering the script to accommodate the locations. Then you make another recce, to cover the same ground again. This time it's on a more down-to-earth basis. You want to know how difficult it will be to shoot in a particular location, and you start thinking about camera angles, lighting equipment, street noises, the availability and cost of local crews, housing, transport, and other depressingly practical matters."

On his second tour of the location sites, in early December, Ross was accompanied by Douglas Slocombe, who had just signed on as cinematographer. Slocombe is an acknowledged master at capturing the aura of period and place on film, and he has won many awards and citations for his contributions to such productions as The Lion in Winter, Travels with My Aunt, and Julia.

Another important member of Ross's second reconnaissance mission was assistant director Ariel Levy, a forceful and witty young man whose unenviable task it would be to maintain some semblance of order amid the inherent chaos of movie-making. During the weeks preceding the start of production, he produced a formidable document known as "the breakdown," which attempted to anticipate and systematize all the requirements for each of the 102 different sets and locations in the film script. For the opening scene in the Budapest hotel suite, as an example, Ariel Levy's breakdown lists the following as props:

Flowers, basket of fruit, an assortment of medicine bottles with labels to read as script, spirit lamp (practical), goose grease in pot or crucible, towel poultice and six repeats, huge sapphire ring, bedside lamp, Nijinsky's and Diaghilev's luggage, tray with remains of a meal, portable samovar, tea glasses, fuel for stove, bills, correspondence, ikons, photo of Nijinsky's mother, ballet sketches, swatches of material, scores, magazines, newspapers.

The breakdown also lists various wardrobe items needed for the scene, down to such minutiae as the snow that would have to show on Diaghilev's overcoat. It goes without saying that everything on the list had to be, or at least look, absolutely authentic—magazines and newspapers of 1912 (they would have to be specially printed), medicines of the period, ditto the luggage, etc. Although the filming of this scene might not occur for months (it actually went before the camera in early April 1979), all the props and costumes that could conceivably be required for it had to be foreseen and procured long in advance.

Along with the breakdown, Ariel Levy had responsibility for compiling the shooting schedule. This went through various transformations in the pre-production period, and would later be updated and amended according to the exigencies of the moment. But as 1978 drew to a close, the essential planning for the film had been set. Shooting was to start the last week of January.

New Year's Day 1979 in London dawned cold and gray, the harbinger of Britain's worst winter in two decades. That morning Leslie Browne flew in from New York. With her arrival, intensive rehearsals could begin in earnest. "The weeks leading up to the start of shooting are always hectic," Ross says, "but these were particularly fraught. George and Leslie were trying to stretch their voices with the help of speech coaches. Carla was working on her pronunciation of English. Bill Griffiths, one of America's best dance coaches, whom we had brought over from New York, was giving class to the dancers every day. And all the time Nora and I were supervising rehearsals, keeping an eye on costume fittings and makeup tests, looking at set designs, and watching progress on the ballet stage at Pinewood."

In The Turning Point, Ross had shot the ballet scenes in a real theater—the Minskoff in New York. This time, in order to gain greater flexibility for camera placement, he decided to film the dance sequences in a studio. "The idea," he explains, "was to approximate theater conditions as closely as possible but still retain the freedom of studio filming." Over a period of several weeks, a Pinewood construction crew erected a facsimile stage platform and wings within the capacious shell of the studio's vast Stage E. A theatrical lighting consultant, David Hersey, was then brought in to light the stage as if it were in a real theater. "David would work all morning at Pinewood, lighting one of the ballet sets," Ross recalls, "and in the afternoon Nora and I would drive out to look it over. We'd usually fiddle around a bit with colors and inten-

sities, and then Douggie [Slocombe] would shoot a test. Once we had achieved what we wanted, the lighting mix would be memorized in a computer. Before shooting started, we knew exactly how the ballet sequences would look on film."

Pinewood Studios, located about eighteen miles west of London in the general vicinity of Heathrow Airport, comprise a sprawling complex of administration buildings, sound stages, dressing rooms, viewing theaters, costume and property warehouses, and machine and carpentry shops—an enormous factory of make-believe set in the open countryside of Middlesex. Here on Thursday, January 25, 1979, at 9:00 A.M.—after fourteen months of nonstop preparation and 43 years of false starts—the filming of Nijinsky at last got under way.

The first nine days were atypical. Nijinsky is essentially a film of drama, not spectacle, yet at the start of production it was spectacle that predominated. Because of the London Festival Ballet's performance schedule, it had been decided to film all the ballet sequences at the very start of production. Seven scenes that would later be scattered throughout the film were lumped together into one dance marathon. Le Spectre de la Rose, with only a pair of dancers involved, came first on the schedule and took up two days. Next on the list was the Carnival Scene from Petrouchka, in which the entire ballet company took part. It was shot in one day, but eventually wound up on the cutting room floor, along with the sequence from Prince Igor. Excerpts from Schéhérazade and Prince Igor were filmed over the following three days, again with the entire company. Two further days took care of L'Après-midi d'un Faune, the Cell Scene from Petrouchka, and parts of Jeux and Le Sacre du Printemps. On the final day of ballet shooting, work was completed on the Spectre and Faune sequences. Except for a few backstage interjections, not a word—indeed, not a sound—was recorded during this period. The dancing was filmed to temporary pre-recorded piano playback. Only much later would a new orchestra soundtrack, under John Lanchbery's musical direction, be recorded in synch to the film.

The camera setups during these first nine days were also atypical. Ordinarily, only one camera is employed on a film set. A scene will be shot from one point of vision, and then reshot from another. But with an entire ballet company on payroll, the luxury of retakes from different camera positions would have been prohibitively expensive. Thus, in filming the ballet sequences, Slocombe had five, sometimes even six, cameras turning simultaneously. He favored very wide apertures "to spread the highlights," and made extensive use of gauzes "for a soft, dreamlike quality."

Each afternoon during the production of a film, a "call report" is distributed to crew and principals, alerting everybody to the following day's schedule. Some notion of the sheer enormity of film-making can be gleaned from a look at the Nijinsky call report for January 29, 1979. The day started at 6:15 A.M., when a fleet of cars began picking up the director, the cinematographer, and the principal actors and dancers at their residences in central London. A half-hour later, the London Festival Ballet's corps de ballet and attendant personnel (some 90 people in all) left London for Pinewood in two private coaches. By 7:00 the principals had arrived at Pinewood, and immediately began dance class under the supervision of Bill Griffiths. This lasted for an hour. As the dancers walked from class to their dressing rooms in another building, they could make out the first wan glimmer of an English winter morning.

The makeup that day was particularly demand-

ing. In retrospect, Kenneth Lintott—chief makeup artist for the film—commented: "It seemed to all of us that the ballet sequences should somehow express the extraordinary effect that the Ballets Russes exerted on those early audiences, and certainly makeup contributed to that effect. But the greasepaint of 1912 blown up on a large movie screen would look old-fashioned, perhaps even ludicrous, so we had to find a level somewhere between what they did in Diaghilev's time and what we consider acceptable now. The contemporary photos give a very clear picture of the atmosphere —the perfume, if you like—of those first performances. We've tried to convey something of what we saw in the pictures. But Petrouchka turned out to be very difficult. Superficially it would seem easy to duplicate Nijinsky's makeup, because the photos of him in the part are so good. But the more I look at those pictures, the more I'm convinced that his appearance wasn't simply a matter of applying paint to the face. That look of utter negation comes from something going on inside the dancer. Try as we did to superimpose the same look on George's face, we never quite achieved it."

Whether these speculations occurred to Kenneth Lintott as he worked against deadline on De la Pena's makeup at 8:00 in the morning of January 29, is perhaps debatable. At any rate, we know from the call report that by 10:00 the entire dancing cast —principals and corps de ballet—were to be costumed, made up, and in place on Stage E for the start of shooting.

At that same hour, Carla Fracci was meeting Alan Barrett for a costume fitting on King Henry's Road in the Primrose Hill section of London. As we have already noted, Barrett had taken over responsibility for the costumes just a few weeks before. By the time he joined the production, costumes for the dance sequences were well in hand, and arrangements had been made with a leading costumier in Rome to supply garments for the crowd scenes. But Barrett had to start from scratch in dressing the principals. "The job was particularly difficult," he says, "because Herbert Ross wanted to use as many original clothes of the period as possible. Fortunately, I knew of a costume house in London run by a brilliant young man who is batty about old clothes and who has an amazing collection of originals. We drew heavily on his stock for the film. Old clothes are usually very fragile, and one is always repairing them. But it's worth the trouble. You simply can't duplicate the materials and the workmanship today. When I took Carla to this costume house for her fitting, she went crazy. I think she would have bought everything in the place if it had been for sale. Of course, not every garment in the film is an original. Sometimes we had costumes made up from old fabrics. Sometimes we settled for new fabrics that had the right period look. But always, in every outfit in the film, there's something original from the period, whether it's headgear or a scarf or merely a piece of jewelry."

Back at Pinewood, the shooting of <u>Petrouchka</u> continued until the lunch break. At 11:45 A.M., a masseuse arrived on set to help any dancers with sore muscles. Buffet tables were set up to feed 170 people. The call report notes: "Only those of the unit who are unable to leave the shooting area may eat their food on set. Everyone else is asked to eat their food in the area provided." Herbert and Nora Ross, along with associate producer Howard Jeffrey, normally had lunch at the studio restaurant. Slocombe and his crew drove off to a local pub, usually with script girl Louise Jaffe. The principal actors lunched picnic-style with Alan Barrett and Kenneth Lintott in their own wing of dressing rooms.

Shooting recommenced at 2:00 and continued—with a tea break at 3:30—until 5:40 P.M. The Rosses and Howard Jeffrey then went off to a viewing theater to watch rushes from the previous day's shooting. Principals and corps de ballet returned to their dressing rooms for removal of makeup. The construction crew was given this reminder in the call report: "On completion of shooting, strike Petrouchka and erect Schéhérazade set ready for 7:00 A.M. Tuesday, 30th January." With luck, the Rosses and company would be back in central London by 9:30 P.M. Their cars would be picking them up again at 6:15 A.M. the next day.

After nine days of shooting at Pinewood, the staged ballet sequences were all on film. There remained a few additional scenes in which dancers from the London Festival Ballet were to appear—scenes showing Maestro Cecchetti's dance classes and the exhausting rehearsals for Le Sacre du Printemps. These were filmed on location over the next three days in the Library and Alabaster Room of a turn-of-the-century ballet school in Tring Park, some 40 miles northwest of London. Again there were ballet classes in the early morning, and again a masseuse in attendance throughout the day. A new recruit for these scenes was a score consultant, Michael Hyatt, who advised Ross on how accurately Ronald Pickup (as Stravinsky playing the piano) and the dancers were conveying the rhythmic complexities of Le Sacre du Printemps. With the completion of the scenes at Tring Park, Ross and his cast could shift gears from ballet to drama. George de la Pena had still to do one important dancing scene—a solo from Fokine's ballet Carnaval at Lady Ripon's house party—but that was two months in the future. Otherwise the emphasis was now to be wholly on acting out a story.

That story, as in all movies, was to be shot in bits and pieces—one scene here, another scene there, according to convenience, location, and the availability of actors. No attempt would be made to perform the script in consecutive order. Only much later, when the film was being edited, would all the bits and pieces be put together in proper continuity.

A vivid illustration of the process is provided at the very beginning of the picture, as Diaghilev arrives at his hotel in Budapest. Here is how the scene opens in the script:

EXTERIOR. HOTEL HUNGARIA — BUDAPEST — NIGHT

It is snowing. A closed, horse-drawn carriage, with its windows tightly closed, is coming to a stop at the door. SERGEI PAVLOVITCH DIAGHILEV, an imposing, handsome, thirty-nine year old, Russian aristocrat, tending to plumpness, a startling streak of white in his hair, steps out of it with a handkerchief pressed over his mouth. The DOORMAN, excited, blows a whistle. FOUR PAGES, each with an umbrella, come tumbling out of the hotel. TWO PORTERS are standing by.

DIAGHILEV (to DRIVER)
Here! Keep that damn horse away from me! God knows what germs it's harboring.

As the Driver climbs down and is paid by Diaghilev, the Page Boys line up in formation, the Doorman looks in the carriage for luggage. There is, to his surprise, only one small rather battered suitcase. The Doorman hands the suitcase to a Page who follows Diaghilev into the hotel.

This brief moment was filmed over a period of three months in three different locations. The carriage entrance was shot in the courtyard of the Palazzo Biscari in Catania, Sicily, on the night of March 16. It was the last of three scenes filmed on location that evening. By the time Ross had taken Diaghilev's arrival from two different camera setups, it was two o'clock in the morning. Alan Bates had been on call since 1:30 in the afternoon. "I still had to shoot a close-up of Diaghilev looking through the carriage window," Ross says, "but it was too late for that, and we decided to put off the close-up until after we got back to Pinewood, where Douggie could light it under controlled conditions." The carriage close-up, using a facsimile of the vehicle filmed in Sicily, was made in Pinewood's Studio K on April 6. The "hotel" itself

was actually the Geological Museum in Budapest, chosen because of its untouched fin-de-siècle entrance foyer, which could readily be turned into a hotel lobby circa 1912. This part of the scene—in which Diaghilev and the page enter through the front door, walk across the lobby, and start up the stairway—was filmed on location in Budapest on the afternoon of February 17. Admittedly, three locations in three countries for a scene lasting less than two minutes is an extreme example of the discontinuity of filming, but it will serve to underscore the difficulty of chronicling the making of a movie.

On February 14, the Nijinsky production moved to Budapest. "The biggest practical problem in making a film like this," says producer Stanley O'Toole, "is the logistics of getting people and equipment from one place to another. When we got ready to leave for Budapest, we found there was so much stuff to take along—especially costumes—that we had to charter two planes, one for people and another for freight. In twenty years in this business, I've never had to do that before. Altogether we had about 110 people on location—the production department; two camera crews; the entire wardrobe department, including several fitters; makeup artists and hairdressers; the design and property people; a publicity staff; a stand-by crew; electricians; a catering unit; and of course the actors and their stand-ins." On arrival in Budapest, this cinematic army dispersed to their hotels—the VIPs to the Hilton, the remainder to the Gellert. For the next two weeks, Hungary would be home.

Budapest's nineteenth-century opera house, a typically Hapsburgian confection of red plush and gilt scrollwork, appears today just as it did in 1912 when Nijinsky first danced there. No location could have been more appropriate as a background for

Romola's introduction to the Ballets Russes and for her ineffectual early attempts at catching the attention of Diaghilev and his star dancer. Five days were spent shooting scenes in various sections of the theater, some with a full complement of extras in period evening dress, others with the principals alone. Work usually had to stop there in mid-afternoon, in order to clear the house for that evening's performance.

Early on a gray Sunday morning, Ross started filming Karsavina's arrival at Budapest's Eastern Station. This required a steam locomotive and carriages of pre-World War I vintage, disguises to cover contemporary signs on the platform, and a crowd of passengers, army recruits, porters, and train mechanics, all milling about in dress appropriate to 1912. "We had only about two hours there," Ross recalls, "and while we were making our shots, trains kept arriving from places like Zagreb and Cracow, ruining our soundtrack."

Two scenes were filmed in turn-of-the-century Budapest cafes, though only one appears in the film. The splendidly overdecorated Cafe Hungaria serves as the First Class dining saloon on the S.S. <u>Avon</u>, carrying the Diaghilev company to South America. Romola is shown descending a short flight of steps and pausing briefly at Nijinsky's table. Though brief, the scene took almost twelve hours to shoot because of the many camera setups required. In the equally redolent Cafe Vorosmarty, Ross filmed the scene in which Karsavina tells Nijinsky the latest news from St. Petersburg. Later he decided to scrap it. "When I saw the rushes, I realized that the concept was wrong. I was shooting the pastry shop instead of the scene. It's a common mistake." One wonders what happened to the "wonderful assortment of cakes and pastries" listed as props in the call report.

On Friday, March 2, the army decamped for Palermo, traveling again in chartered aircraft. After a weekend at leisure in the antique splendor of the Hotel Villa Igiea, cast and crew resumed work on Monday morning. During the next two weeks in Sicily, the <u>Nijinsky</u> company was to film scenes set in Budapest, Greece, Paris and the Riviera, Italy, and South America. A gorgeously decorated Palermo town house, the Palazzo Ganci, served as background for Karsavina's arrival in Buenos Aires and for a scene between Nijinsky and Baron de Gunzburg in the latter's library (presumably in Paris). Some idea of the logistics involved in location work can be gained from the list of vehicles parked outside the Palazzo Ganci:

2 camera vans
3 electrical trucks
1 grips truck
1 props truck
1 W.C. truck
2 catering trucks
1 truck for tables and chairs
4 trailers for the principals and director
1 trailer for makeup and hairdressing
2 generators
1 truck for horses
1 carriage
1 coach for extras

One day each was spent at Segesta and Seli-nunte, the Greek temple sites. In addition to flocks of sheep and a miscellany of vintage picnic gear, Diaghilev's carriage was on hand for these shots. Ross's researches had satisfied him that horse-drawn vehicles were still the principal means of transport in Europe and South America prior to World War I, and he decided to dispense with automobiles of the period as being "too quaint and too comical." A few vintage cars do appear in a scene shot in Monte Carlo, but the principals themselves all ride about in carriages. On the other hand, the telephone is very much in evidence, as is electric lighting.

A third day outdoors was spent at Mondello Beach, near Palermo, filming the opening part of

Diaghilev's and Nijinsky's last scene together. Ross had been lucky. For three days in a row—while he was on exterior locations—the skies remained blue, considerately sustaining Sicily's claims to an early spring.

After a week in Palermo, the unit moved to Catania, on Sicily's eastern shore. Here, in the late-nineteenth-century Teatro Bellini, Ross filmed interior theater scenes to complement the ballet sequences already filmed at Pinewood. It was not enough just to show the ballets being performed on stage. There had as well to be a suitably attired audience watching them. During Le Sacre the spectators vent their outrage by throwing crumpled programs at the stage. Pursuing authenticity to the last iota, facsimiles of the 1913 "Programme Officiel des Ballets Russes" were distributed to everyone in the audience. Fokine's final rupture with Diaghilev was also filmed in this handsome opera house, the two actors playing the scene in an empty auditorium, while on stage an orchestra of 47 musicians—in effect, the Teatro Bellini's regular pit orchestra—rehearsed L'Après-midi d'un Faune.

Two museum-like town houses in Catania were pressed into service for a variety of scenes. In the Palazzo del Toscano, Ross staged the lavish Monte

Carlo party at which Romola is introduced to Diaghilev. Some 275 people were on hand there during two days of shooting. In addition to the entire production unit from London, plus ancillary Italian electricians and grips, there were a crowd of 89 guests and a six-piece "palm court" orchestra, all dressed in period costumes. We have already seen how the Palazzo Biscari figured in Diaghilev's arrival at the Hotel Hungaria early in the script. Other parts of this house served for Stravinsky's music room, where Diaghilev and Nijinsky first hear Le Sacre, and for the dressing room in Buenos Aires, where Nijinsky receives the cablegram dismissing him from the Ballets Russes. Finally, Catania's eighteenth-century cathedral did duty as the background for the wedding scene in Buenos Aires.

After one last free day in Sicily (an outing was organized to nearby Mount Etna), the Nijinsky unit flew off to Monte Carlo. Here the troops were quartered—according to fame and fortune—in a half-dozen billets, from the super-luxe Hotel de Paris down to the lowly Hotel d'Europe. Monte Carlo was vintage Diaghilev-land. The Ballets Russes had been headquartered there for many years, and the Hotel de Paris was as much of a home as Diaghilev ever had after leaving Russia.

Work began on March 22 with a day of outdoor shooting in the Jardin Exotique to film a scene between Romola and the dancer Adolph Bolm (it was later deleted). For Leslie Browne the day started at 5:45 A.M., when a car drove her from the Hotel de Paris to an improvised makeup room in the Hall du Centenaire. She was on location by 7:30 and worked until 4:00 in the afternoon. Midway through the day's shooting, she felt ill. A doctor who attended her ascribed the malaise to fatigue. Fortunately, she would not be on call again until March 27. There was plenty of time to rest.

Ross continued to work at top speed with the rest of the cast. One day was spent filming a scene between Karsavina and Nijinsky in the lobby of the Hotel de Paris. Another day was spent across the road in the Monte Carlo Opera House filming the "dress parade" for Fokine's ballet <u>Le Dieu Bleu</u>. And a third day was spent shooting some fairly brief backstage scenes at the opera house in Nice, a few miles down the coast.

Leslie Browne was back on location the morning of March 27 to film Romola's arrival in Monte Carlo. The station in Monte Carlo has been totally modernized in recent years, but in nearby Menton the director had discovered an old market building that could be transformed with a few vintage signs and props into a reasonable facsimile of a nineteenth-century railroad shed. The train arrival itself was filmed in the actual Menton station; then the unit moved to the market building to shoot Romola as she comes out of the station and rides off in a carriage with her friend Magda. Throughout the day, Leslie Browne kept complaining of the cold. "She was shivering in the Menton station," Ross

recalls, "but we finally got the scene on film. After that we moved to the market building and completed one shot. I was ready to do another when Leslie came over to me and said, 'Herbert, I'm sick. I don't think I can work any more.' We wrapped the day's shooting then, and returned to Monte Carlo.

"Back in Catania, I had noticed that something was wrong. Leslie seemed to be holding back. I thought it was emotional. To overcome what struck me as her self-indulgence, I began pushing Leslie very hard. I couldn't understand why she was constantly tired, why she had bags under her eyes, why she couldn't stand while we were lighting her. But it was only in Menton that I started to get really worried."

The next day an outbreak of bad weather gave the actress a respite. Instead of the scene she was scheduled to play, riding in a carriage along the Monte Carlo Esplanade with Magda, Ross filmed a "cover shot" inside the opera house involving only Diaghilev, Astruc, and Fokine. Leslie Browne's respite was short-lived. That night she was rushed to Princess Grace Hospital for an emergency operation. Her appendix had ruptured and she was discovered to be suffering from acute peritonitis.

Shooting was cancelled while Herbert and Nora waited at the hospital for news of Leslie's condition. By midday she was pronounced out of danger, but the doctors warned that the actress could not resume work for at least a month. The Rosses swallowed hard, called a meeting with the production staff, and began formulating alternative plans to keep the picture going.

According to their original schedule, Ross and the company were to have continued working in and around Monte Carlo for another six days. They were then to have gone to Genoa for three days of shooting on board the S.S. Iripina before returning to England. The Iripina was an old Italian steamer, still in service, which had been selected as a location for all the deck and quayside scenes. It had now to be scuttled, at least as far as the film was concerned. Since all the shipboard scenes involved Romola, there was no point in going to Genoa. Somehow or other, another nautical location would have to be found later on. A few scenes could still be filmed on the Riviera, however.

On March 30, the sun shone again. The Esplanade scene, which had been cancelled because of bad weather two days earlier, was now filmed as a long shot, with a double as Romola—her back to the camera—sitting beside Magda in the carriage. At the same time, background shots of the Esplanade were also filmed. Later on, in the Pinewood studio, close-ups of Leslie Browne would be made in the same carriage against a back-projection of the Esplanade. Two days of filming in the Rothschild villa at St. Jean-Cap Ferrat com-

pleted the Riviera location work. Kid-glove treatment was ordained here. "In view of the precious and fragile nature of the location we are working in," the call report warned, "everyone is urged to take utmost care when handling and placing equipment." Two hotel scenes—one between Karsavina and Nijinsky, the other between Karsavina and Diaghilev—were filmed in fragile Rothschild bedrooms. Then the unit packed up and flew back to London, leaving Leslie Browne and her mother to follow in due course.

By dint of working overtime, John Blezard and his crew had several sets at Pinewood ready for shooting when the Nijinsky unit arrived there on April 3. Over the next three weeks Ross was to complete all the remaining scenes in which Leslie Browne's presence was not required. Much of the action from this point on was to be played in hotel rooms, set variously in Budapest (for the opening scene between Diaghilev and Nijinsky), Athens, Berlin, Paris, Venice, Buenos Aires, and Genoa. These endless hotel sequences were a constant challenge to Blezard's ingenuity. "Usually," he says, "you analyze the characters in a script and try to design sets that somehow reflect their personalities. It wasn't possible to do that in this film. Hotel suites, shipboard cabins, train compartments are by their very nature impersonal." Blezard attempted to get around the monotony by using varying styles of decor and allowing the camera to glimpse different backgrounds beyond the window curtains—a street corner in Budapest, a palm-lined boulevard in Buenos Aires, a blue bay in Genoa. "One had always to be thinking," Ross adds, "of color, texture, wallpaper, the quality of light. These were the details that allowed us to convey a suggestion of place. In addition, we were continually having to invent activities for the characters. Because Diaghilev and Nijinsky lived in hotels all the time, everything was done for them. So one had to create activities—taking baths, getting shaved, having shoes shined, that sort of thing. Otherwise they would have been simply sitting around talking, as in a play."

The routine of shooting scenes in the facsimile hotel suites and shipboard interiors at Pinewood was interrupted by two sorties to stately homes within driving distance of the studio. Lady Ripon's outdoor party was staged at Luton Hoo, an imposing edifice rebuilt and redecorated in the early twentieth century. At the back of this mansion, a porticoed terrace looks out over an expanse of formal lawns and gardens. Here a stage had been set up for Nijinsky's solo performance, and near it was a grand piano to provide the accompaniment. The weather, still bitterly cold and depressingly gray, made this a particularly hard scene to shoot. Some fifty extras were on hand to play Lady Ripon's guests, the women in Edwardian silk gowns, feather boas, and diamond tiaras—splendid costumes to look at, but of scant comfort on a damp day with near-freezing temperatures. Inside the house, which was kept relatively warm with a battery of portable heaters, George de la Pena worked at the barre to the accompaniment of Bill Griffiths's snapping fingers. De la Pena's costume for the role of Harlequin in Fokine's ballet Carnaval was the least protective of all, and he would wait until the last minute before going out to brave the cold.

The other stately home, Waddesdon Manor, once a Rothschild residence and now owned by the National Trust, was built a century ago in the style of a French chateau. Waddesdon was used to film the party scene that immediately follows the performance of Le Spectre de la Rose in Budapest. Altogether some 200 people—including ex-

tras to play guests, waiters, and footmen—were deployed there, plus all the attendant coaches, trucks, trailers, and catering facilities to transport, equip, shelter, and feed them. "It's a very short scene," Ross observes, "and you might well ask why we went to all that trouble. I did it because I thought it important to establish early in the movie the atmosphere of high society in which Diaghilev and his chief dancers moved. Waddesdon provided the perfect background."

The scene there was filmed on April 24. Two days later, Leslie Browne returned to Pinewood for a costume fitting, and on the 27th she was back on set to film the scene where Romola encounters Nijinsky on a stairway of the S.S. Avon. That same

day saw the arrival of Cecil Beaton, doyen of British portrait photographers. Over the course of the next two weeks, Sir Cecil was to photograph every member of the cast in costume.

With Leslie Browne in action again, Ross could tackle the major dramatic scenes between Romola and Vaslav. Before they were over, everyone—including the director—had been totally drained.

During a break in filming, Ross explained why: "The picture has taken a turn I didn't foresee. The last scenes are much more difficult than I had anticipated, and they go deeper than anything I've previously attempted."

By the end of the first week in May, there remained only the deck and quayside scenes to be filmed. The S.S. Iripina had long ago left port, and anyway the expense of transporting the entire unit to Genoa for two days on location would have been prohibitively costly. Blezard had come up with a less daunting alternative. At Shepperton, another large film studio on the outskirts of London, work had just concluded on a picture called S.O.S. Titanic. A huge replica of the doomed ship's hull and decks, erected within one of the sound stages at Shepperton, was still intact. This formidable set was taken over by the Nijinsky production. What had previously been the S.S. Titanic would now become the S.S. Avon (for the outward voyage to South America) and the S.S. Regina Elena (for the disembarkation at Genoa). First, however, the set had to be dismantled and put up on new scaffolding outdoors, because Ross wanted to photograph real sky and real clouds. It was a considerable undertaking, and the Shepperton construction crew managed to finish the job just in time.

In effect, the shipboard and gangplank scenes at Shepperton completed the filming of Nijinsky. A few loose ends remained to be tied up at Pinewood—various inserts for scenes that had already been shot on location—but otherwise the long march was over. On May 14, the eve of the final day at Pinewood, a farewell party was thrown for the cast and crew. It seemed a subdued affair. The Nijinsky "family" had been together for nearly five months. Nobody appeared overjoyed at the prospect of its breaking up.

By the end of the week, Herbert and Nora Ross had departed for New York and, later, Hollywood. Months of painstaking labor lay ahead—cutting, editing, re-recording, re-processing, titling—all the myriad operations that would turn 474,555 feet of 35-mm color stock into a coherent and compelling motion picture. But the actual process of creation had run its course. Nijinsky would stand or fall according to what lay within those 237 cans of unedited film.

The day before he left London, Ross allowed himself a moment of retrospection. "I think we were all remarkably stupid and remarkably brave to go ahead on this picture, particularly when you think of all the talented people in the past who tried to do it and failed. The film was an enormous and complex and challenging piece of work—for me, for Nora, for the cast, for literally everybody involved with it. We all approached it with a genuine sense of mission. This was much more to us than just another movie."

NIJINSKY

THE FILM

BUDAPEST
Hotel Hungaria, 1912

Diaghilev and Nijinsky
are touring with the Ballets Russes

Tamara Karsavina arrives in Budapest

Romola de Pulsky attends
a Ballets Russes performance

The company in daily class with
ballet master Enrico Cecchetti (far left)

Nijinsky in costume
for <u>Le Spectre de la Rose</u>

Original poster art by Cocteau
for <u>Le Spectre de la Rose</u>

Nijinsky and Karsavina dance
in <u>Le Spectre de la Rose</u>

Nijinsky and Karsavina perform <u>Le Spectre</u>

Photo session
for <u>The Firebird</u> costumes (right)

Romola begins her pursuit of Nijinsky

Diaghilev and Nijinsky in Greece, 1912

Nijinsky finds inspiration
for L'Après-midi d'un Faune

Nijinsky begins to choreograph <u>Faune</u>

Bakst's costume drawing
of Nijinsky as the faun

Bakst's costume designs for <u>Le Dieu Bleu</u>

Nijinsky dances as
the Golden Slave in <u>Schéhérazade</u>…

...with Karsavina as Zobeida

MONTE CARLO, 1912

Romola continues her pursuit of Nijinsky
by trying to join the Ballets Russes

PARIS, May 29, 1912

Nijinsky dances
in the first ballet he created,
L'Après-midi d'un Faune

The controversy created by <u>Faune</u>
served chiefly to increase the fame
of the Ballets Russes

PARIS, May 15, 1913

<u>Jeux</u>, the second ballet choreographed
and danced by Nijinsky

ENGLAND, 1913

Nijinsky dances as Harlequin
from Carnaval at a party given by
the Marchioness of Ripon

Romola auditions for Cecchetti

Romola takes class with
the Ballets Russes

Nijinsky begins to create
Le Sacre du Printemps
with the composer Stravinsky

PARIS, May 29, 1913

The riotous first performance
of <u>Le Sacre</u> detonated one of the
loudest explosions in theatrical history:
the critics as well as the audience
responded violently

ITALY, 1913

After the failure of <u>Le Sacre</u>,
Nijinsky and Diaghilev take a holiday in Italy

New tensions develop in their relationship,
and it is decided that Diaghilev
will stay in Europe while Nijinsky goes
to Buenos Aires on tour with the company

On board the S.S. <u>AVON</u>, 1913

Estranged from Diaghilev,
Nijinsky finds solace in the arms of Romola

BUENOS AIRES, September 10, 1913

Nijinsky and Romola are married

Diaghilev learns of the marriage
and sends a cable to Nijinsky
terminating his affiliation with
the Ballets Russes

Nijinsky in <u>Petrouchka</u>

The company returns to Europe,
and Karsavina is unsuccessful
in her attempt to mend the rift.
Romola asks Diaghilev to take Nijinsky back,
but is rebuffed.

Vaslav Nijinsky's last performance was in 1917.
He would spend the next thirty-three years in mental institutions.
He died in 1950.